SPATIAL ECONOMETRICS: METHODS AND MODELS

STUDIES IN OPERATIONAL REGIONAL SCIENCE

Folmer, H., Regional Economic Policy. 1986. ISBN 90-247-3308-1.

Brouwer, F., Integrated Environmental Modelling: Design and Tools. 1987. ISBN 90-247-3519-X.

Toyomane, N., Multiregional Input–Output Models in Long-Run Simulation. 1988. ISBN 90-247-3679-X.

Spatial Econometrics:
Methods and Models

by

Luc Anselin

Departments of Geography and Economics,
University of California, Santa Barbara

KLUWER ACADEMIC PUBLISHERS
DORDRECHT / BOSTON / LONDON

Library of Congress Cataloging in Publication Data

Anselin, Luc, 1953-
    Spatial econometrics : methods and models / by Luc Anselin.
        p.   cm. -- (Studies in operational regional science)
    Includes index.
    ISBN 9024737354
    1. Space in economics--Econometric models. 2. Regional economics-
    -Econometric models.   I. Title.   II. Series.
    HB199.A49 1988
    330'.028--dc19   88-16971
                                                                   CIP

ISBN 90-247-3735-4

Published by Kluwer Academic Publishers,
P.O. Box 17, 3300 AA Dordrecht, The Netherlands.

Kluwer Academic Publishers incorporates
the publishing programmes of
D. Reidel, Martinus Nijhoff, Dr W. Junk and MTP Press.

Sold and distributed in the U.S.A. and Canada
by Kluwer Academic Publishers,
101 Philip Drive, Norwell, MA 02061, U.S.A.

In all other countries, sold and distributed
by Kluwer Academic Publishers Group,
P.O. Box 322, 3300 AH Dordrecht, The Netherlands.

printed on acid free paper

02-0692-200 ts

to E.T.

TABLE OF CONTENTS

LIST OF TABLES

LIST OF FIGURES

PREFACE

Spatial econometrics deals with spatial dependence and spatial heterogeneity, critical aspects of the data used by regional scientists. These characteristics may cause standard econometric techniques to become inappropriate. In this book, I combine several recent research results to construct a comprehensive approach to the incorporation of spatial effects in econometrics. My primary focus is to demonstrate how these spatial effects can be considered as special cases of general frameworks in standard econometrics, and to outline how they necessitate a separate set of methods and techniques, encompassed within the field of *spatial econometrics*.

My viewpoint differs from that taken in the discussion of spatial autocorrelation in spatial statistics – e.g., most recently by Cliff and Ord (1981) and Upton and Fingleton (1985) – in that I am mostly concerned with the relevance of spatial effects on model specification, estimation and other inference, in what I call a model–driven approach, as opposed to a data–driven approach in spatial statistics. I attempt to combine a rigorous econometric perspective with a comprehensive treatment of methodological issues in spatial analysis.

Although I started working on these issues almost ten years ago, as part of my doctoral dissertation at Cornell, most of the material in this book is much more recent. Some of it appeared earlier in various forms in a number of journal articles. However, much of the book consists of so far unpublished methods and findings, and the bulk of the empirical examples are new. By combining a fairly technical treatment of tests and estimators with an extensive set of illustrations and practical considerations, I hope I have achieved a mix which is of interest both to researchers in regional science and spatial analysis as well as to applied econometricians. The breadth of the topics considered should also make the book appropriate as a graduate level text in spatial statistics, econometrics and regional analysis.

In the development of the research behind this book, and in the writing of previous versions of various parts of the manuscript, I have benefitted greatly from the comments, suggestions and encouragement from a number of people, and especially from Walter Isard, Art Getis, Reg Golledge, Geoff Hewings and Peter Nijkamp. Much of the research was supported by grants SES 83–09008 and SES 86–00465 from the National Science Foundation, and part of the empirical work stems from a grant by the Ohio Board of Regents Urban University Research Program and a University of California at Santa Barbara Senate Research Grant. The Regional Science Association granted permission to use part of the materials from my article on "Specification Tests on the Structure of Interaction in Spatial Econometric Models," which appeared in volume 54, 1984, of the *Papers, Regional Science Association.*

Susan Kennedy and Serge Rey read the manuscript and provided comments from a graduate student perspective. The graphics were produced by David Lawson and Serge Rey, and other general research assistance was provided at various points in time by Ayse Can, Mitchell Glasser, Steve Mikusa and Serge Rey.

The book is dedicated to Emily, who served as general editor and graphic illustrator, but who also put up with my one−track mind during the past months.

Santa Barbara, January 1988

CHAPTER 1

INTRODUCTION

The importance of space as the fundamental concept underlying the essence of regional science is unquestioned. Since the early growth of the field in the late 1950s, a large number of spatial theories and operational models have been developed which have gradually disseminated into the practice of urban and regional policy and analysis. However, this theoretical contribution has not been matched by a similar advance in the methodology for the econometric analysis of data observed in space, i.e., for cross−sections of regions at one or more points in time.

Aggregate spatial data are characterized by dependence (spatial autocorrelation) and heterogeneity (spatial structure). These spatial effects are important in applied econometric analysis, in that they may invalidate certain standard methodological results, demand adaptations to others, and, in some contexts, necessitate the development of a specialized set of techniques. These issues are typically ignored by traditional econometrics and have been encompassed by the separate field of *spatial econometrics*.

The problem of spatial dependence, and particularly spatial autocorrelation, has received substantial attention from a statistical perspective, e.g., most recently in Cliff and Ord (1981) and Upton and Fingleton (1985). However, even in these highly specialized texts, the treatment of spatial effects from an econometric viewpoint is rather cursory, and limited to simple problems in the linear regression model.

In this book, I attempt to go beyond this state of affairs by outlining a comprehensive approach to dealing with spatial effects from an econometric perspective. The main objectives are to illustrate how spatial effects can be viewed as special cases of general issues of model specification and estimation in econometrics, to assess the limitations of the standard techniques in a spatial context, and to suggest the required alternative methods.

Many results that have appeared in a variety of sources are brought together. They are combined with several new techniques in a framework that includes spatial dependence as well as spatial heterogeneity, in a purely cross−sectional as well as in a space−time context.

The book is organized around three main issues, and is correspondingly divided into three parts, one dealing with *Foundations for the Econometric Analysis of Spatial Processes*, one with *Estimation and Hypothesis Testing*, and one with *Model Validation*.

In the first part, *Foundations for the Econometric Analysis of Spatial Processes*, I provide a more extensive motivation for the need for a separate field of spatial econometrics, and introduce the formal probabilistic framework for the analysis of spatial processes. Part One consists of four chapters.

1

In Chapter two, *The Scope of Spatial Econometrics*, I further outline the special characteristics of the field, and situate it within the larger discipline of regional science. I also formally define the two aspects of spatial effects, *spatial dependence* and *spatial heterogeneity*, that merit special attention from a methodological viewpoint.

Chapter three is concerned with *The Formal Expression of Spatial Effects*; in other words, with the meaning of *space* in spatial econometrics. In particular, the various ways in which the structure of spatial dependence can be reflected in general spatial weight matrices are outlined, and some problems associated with this approach are discussed.

In Chapter four, I present *A Typology of Spatial Econometric Models*, which underlies the treatment of methods in the later chapters. Two separate taxonomies are outlined, one pertaining to spatial linear regression models for cross—section data, the other to spatial linear regression models for space—time data.

Chapter five, the last chapter in Part One, deals with *Spatial Stochastic Processes: Terminology and General Properties*. In it, I outline some fundamental properties of spatial stochastic processes and their implications for econometric analysis. I also discuss in some detail the formal aspects of an asymptotic approach to the analysis of spatial processes, and evaluate the implications of these concepts for empirical practice in regional science.

The second and also the largest part in the book is concerned with *Estimation and Hypothesis Testing*. It consists of seven chapters: Chapters six through eight deal with spatial dependence, Chapter nine focuses on spatial heterogeneity, Chapter ten introduces space—time models, Chapter eleven outlines some special methodological problems, and in Chapter twelve an extensive empirical illustration is provided for the various estimators and tests.

The Maximum Likelihood Approach to Spatial Process Models is the most familiar framework for carrying out estimation and hypothesis testing in spatial econometrics. In Chapter six, this approach is discussed in detail, with special reference to a general specification which combines spatial dependence and spatial heterogeneity. After a discussion of the limitations of the standard ordinary least squares estimator for this type of model, maximum likelihood estimation is outlined. This is followed by a comprehensive treatment of hypothesis testing within the maximum likelihood framework, which consists of Wald tests, Likelihood Ratio tests and Lagrange Multiplier tests.

Chapter seven considers *Alternative Approaches to Inference in Spatial Process Models*. Three viewpoints in particular are discussed: instrumental variable estimation, Bayesian techniques and robust methods.

In Chapter eight, I focus more specifically on *Spatial Dependence in Regression Error Terms*. After an overview of traditional approaches, I introduce a number of new tests for residual spatial autocorrelation in non—standard situations. In addition, several estimation procedures are outlined and evaluated, with a particular emphasis on issues of robustness.

In Chapter nine, I switch to the topic of *Spatial Heterogeneity*. After a brief discussion of general aspects of this problem, I outline several new tests for heterogeneity in the presence of spatial dependence. I also discuss a number of econometric aspects of the spatial expansion method. The chapter is concluded with a review of several other forms in which spatial heterogeneity can be incorporated into econometric models.

The tenth chapter deals with *Models in Space and Time*. Specific attention is paid to the incorporation of spatial dependence into seemingly unrelated regressions and into error component models. Maximum likelihood estimation is discussed, and some new tests for the presence of spatial autocorrelation are introduced. Also considered are some aspects of the implementation of simultaneous models in space–time.

In Chapter eleven, some *Problem Areas in Estimation and Testing for Spatial Process Models* are considered more closely. Specifically, the pre–test problem in the context of the estimation of spatial models is introduced, the importance of edge effects is discussed, and some brief comments are formulated about the specification of spatial weight matrices and the relevance of sample size in empirical spatial econometrics.

Part Two is concluded with Chapter twelve, *Operational Issues and Empirical Applications*. A number of implementation issues related to estimation by means of maximum likelihood methods are discussed, and a brief review is given of the availability (or, lack of availability) of spatial econometric software. This is followed by a series of empirical examples of various models and methods, both for cross–sectional data as well as for space–time data. The cross–sectional examples are organized around a simple spatial model for neighborhood crime. A total of eight different modeling situations are considered, dealing with various combinations of spatial dependence and spatial heterogeneity. The space–time examples illustrate a simple multiregional Phillips–curve model of county wage changes. It is used to demonstrate four different modeling situations in seemingly unrelated regressions and error component models.

The third part of the book focuses on issues of *Model Validation*. In consists of two chapters. In Chapter thirteen, *Model Validation and Specification Tests in Spatial Econometric Models*, the general problem of model validation is considered more closely, and two types of specification tests are discussed in particular. One consists of the tests on spatial common factors, the other of tests on non–nested hypotheses. Chapter fourteen deals with *Model Selection in Spatial Econometric Models*, and briefly surveys various measures of fit, information based indicators, Bayesian approaches, and heuristics. It is concluded with a summary of practical implications of model validation in spatial econometrics.

Finally, some conclusions and suggested directions for future research in spatial econometrics are formulated in chapter fifteen.

PART I

FOUNDATIONS FOR THE ECONOMETRIC ANALYSIS

OF SPATIAL PROCESSES

CHAPTER 2

THE SCOPE OF SPATIAL ECONOMETRICS

In this first chapter in Part I, I outline the scope of the book in more detail, and describe the meaning of some important terms in an informal manner.

The chapter consists of two sections. In the first, I discuss the relevance of a separate field of spatial econometrics for applied regional science. I also briefly address the distinction between spatial statistics and spatial econometrics. In the second section, I consider the notion of spatial effects in more precise terms, and outline the motivation for taking into account spatial dependence and spatial heterogeneity in empirical regional analysis.

2.1. Spatial Econometrics and Regional Science.

In the practice of regional science, one attempts to address issues and problems faced by cities and regions by drawing on a wealth of theoretical formulations about human spatial behavior. In order to achieve this in an operational context, the theories need to be translated and transformed from abstract formulations to implementable models. This implies that concepts and relations need to be expressed in a formal mathematical specification, that variables need to be given meaning in the context of available data and measurements, and that estimation, hypothesis testing and prediction need to be carried out. This is typically based on a statistical or econometric methodology.

The collection of techniques that deal with the peculiarities caused by space in the statistical analysis of regional science models is considered to be the domain of spatial econometrics. Before dealing more formally with what is meant by space in this context, I first briefly discuss the delineation of the discipline itself.

The term spatial econometrics was coined by Jean Paelinck in the early 1970s to designate a growing body of the regional science literature that dealt primarily with estimation and testing problems encountered in the implementation of multiregional econometric models. In their book *Spatial Econometrics*, Paelinck and Klaassen outline five characteristics of the field in terms of the types of issues considered: [1]

 — the role of spatial interdependence in spatial models
 — the asymmetry in spatial relations
 — the importance of explanatory factors located in other spaces
 — differentiation between ex post and ex ante interaction
 — explicit modeling of space.

Since this initial statement, the term spatial econometrics has found broad acceptance and has come to encompass a wide variety of methods and models.

Nevertheless, the need for a separate field is sometimes questioned, as it is not always clear what distinguishes spatial econometrics from non—spatial (standard) econometrics on the one hand, and from spatial statistics on the other hand. I will now consider these issues in turn.

2.1.1. Spatial Econometrics and Standard Econometrics.

I first consider the distinction between spatial econometrics and traditional econometrics. This can be approached from two main viewpoints. In one, the focus is on the subject matter. Accordingly, all statistical analyses of economic models in regional science could be considered to fall within the scope of spatial econometrics. This classification would be similar to the distinction between spatial economics and non—spatial economics advanced by Isard (1956) some thirty years ago. In this sense, activities such as the estimation of spatial interaction models, the statistical analysis of urban density functions, and the empirical implementation of regional econometric models could all be considered as falling within the scope of spatial econometrics. However, since most of these analyses can be (and are) carried out using standard econometric techniques, this distinction is not very useful. Moreover, much applied statistical analysis of the models in question tends to ignore specific spatial aspects, so that it could hardly be considered as constituting a separate methodology. [2]

Another, more relevant way of viewing the field is much narrower, and focuses on the specific spatial aspects of data and models in regional science that preclude a straightforward application of standard econometric methods. I will call these particular aspects *spatial effects* and distinguish between two general types: *spatial dependence* and *spatial heterogeneity*.

Of the two, spatial dependence or spatial autocorrelation is best known and acknowledged most often, particularly following the pathbreaking work of Cliff and Ord (1973). It is generally taken to mean the lack of independence which is often present among observations in cross—sectional data sets. This dependence can be considered to lie at the core of the disciplines of regional science and geography, as expressed in Tobler's (1979) *first law of geography*, in which "everything is related to everything else, but near things are more related than distant things." In this sense, spatial dependence is determined by a notion of relative space or relative location, which emphasizes the effect of distance. When the notion of space is extended beyond the strict Euclidean sense, to include Isard's (1969) general space (including policy space, inter—personal distance, social networks, etc.), it is clear that spatial dependence is a phenomenon with a wide range of application in the social sciences. [3]

Spatial dependence can be caused by a variety of measurement problems often encountered in applied work. Examples of these are the arbitrary delineation of spatial units of observation (e.g., census tracts, county boundaries), problems of spatial aggregation, and, most importantly, the presence of spatial externalities and spill—over effects. In addition, and quite separate from this measurement issue, the inherent spatial organization and spatial structure of phenomena will tend to generate complex patterns of interaction and dependencies which are of interest in and of themselves. The resulting models of spatial flows, spatial pattern, spatial structure and spatial processes implicitly or explicitly incorporate elements of spatial dependence. [4]

At first sight, spatial dependence may seem similar to the more familiar time—wise dependence encountered in econometric tests for serial correlation, in distributed lag models and other time series analyses. However, this is only partially the case. Upon closer examination, it often follows that standard econometric results from time series analysis do not carry over in a straightforward way to spatial dependence in cross—sectional samples. This is primarily a result of the multidirectional nature of dependence in space, which, as opposed to a clear one—directional situation in time, precludes the application of many simplifying results and necessitates the use of a different methodological framework.

These problems are largely ignored in the standard econometric literature, probably due to its predominant emphasis on dynamic phenomena and time series data. For example, a discussion of spatial effects is totally absent in most of the more popular econometrics textbooks, such as Maddala (1977) and Pindyck and Rubinfeld (1981), as well as in more advanced treatments, such as Vinod and Ullah (1981) and Amemiya (1985). When spatial dependence is mentioned at all, its treatment is usually limited to a brief remark. For example, Kennedy (1985) points to its existence as a possible cause for autocorrelation in regression error terms, and in the encyclopedic text by Judge et al. (1985) it is listed as one of the motivations for the use of seemingly unrelated regression models.

Exceptions are Kmenta (1971) and Johnston (1984). The former considers spatial dependence as a problem in the pooling of cross—sectional and time series data, and states (Kmenta 1971, p. 512): "In many circumstances the most questionable assumption ... is that the cross—sectional units are mutually independent. For instance, when the cross—sectional units are geographical regions with arbitrarily drawn boundaries — such as the states of the United States — we would not expect this assumption to be well satisfied." In Johnston (1984) spatial autocorrelation is briefly outlined as one of the forms of error dependence in the linear regression model. However, in general, spatial effects are not taken into account, even in the methods recently developed for the analysis of panel data, where the focus is on the time dimension, and potential dependence across cross—sectional units is typically assumed away. [5]

The second type of spatial effect, spatial heterogeneity, is related to the lack of stability over space of the behavioral or other relationships under study. More precisely, this implies that functional forms and parameters vary with location and are not homogeneous throughout the data set. For instance, this is likely to occur in econometric models estimated on a cross—sectional data set of dissimilar spatial units, such as rich regions in the north and poor regions in the south. To the extent that the heterogeneity can be directly related to location in space (such as north and south in the example), I will designate it by the term spatial heterogeneity. [6]

In contrast to the spatial dependence case, the problems caused by spatial heterogeneity can for the most part be solved by means of standard econometric techniques. Specifically, methods that pertain to varying parameters, random coefficients and structural instability can easily be adapted to take into account such variation over space. However, in several instances, theoretical knowledge of the spatial structure inherent in the data may lead to more efficient procedures. Also, the complex interaction which results from spatial structure and spatial flows may generate dependence in combination with heterogeneity. In such a situation, the problem of distinguishing between spatial dependence and spatial heterogeneity is highly complex. [7] In those instances, the tools provided by standard econometrics are inadequate and a specific spatial econometric approach is necessary.

In summary, and for definitional purposes, I will consider the field of spatial econometrics to consist of those methods and techniques that, based on a formal representation of the structure of spatial dependence and spatial heterogeneity, provide the means to carry out the proper specification, estimation, hypothesis testing, and prediction for models in regional science.

2.1.2. Spatial Econometrics and Spatial Statistics.

The distinction between spatial econometrics and spatial statistics is less straightforward, and methods tend to be categorized as belonging to one or the other field depending on the personal preference of the researcher.

One possible categorization can be extracted from an interchange between Haining (1986b) and Anselin (1986b) in a recent issue of the *Journal of Regional Science*. In that discussion, the primary distinguishing characteristic seems to be the data-driven orientation of most writings in spatial statistics and the model-driven approach in spatial econometrics. This can be related to a primary concern of many studies in the geographical literature with extracting, identifying and estimating spatial structure or spatial processes from a given data set. However, a more practical classification would probably be based on whether the authors in question refer to their work as spatial statistics or spatial econometrics.

Representative examples of the spatial statistical literature are most of the writings of Cliff and Ord (1973, 1981), Bartlett (1975), Bennett (1979), Griffith (1980, 1987), Ripley (1981), Wartenberg (1985), and Upton and Fingleton (1985), and the bulk of the papers included in the edited volumes by Wrigley (1979) and Gaile and Wilmott (1984).

Spatial econometric writings tend to start from a particular theory or model and focus on the problems of estimation, specification and testing when spatial effects are present. Illustrative examples are the work of Hordijk (1974, 1979), Hordijk and Paelinck (1976), Hordijk and Nijkamp (1977), Arora and Brown (1979), Blommestein (1983), Blommestein and Nijkamp (1986), Anselin (1980, 1987a), Foster and Gorr (1986), and most of the papers included in Bartels and Ketellapper (1979), Bahrenberg, Fischer and Nijkamp (1984), and Nijkamp, Leitner and Wrigley (1985).

Also, spatial econometrics typically deals with models related to regional and urban economics, whereas a substantial body of the spatial statistical literature is primarily focused on physical phenomena in biology and geology (e.g., point pattern analysis, kriging) and thus less directly relevant to issues studied in regional science.

This distinction between the two fields is far from clear and several methodological issues are equally considered in both. In order to avoid terminological confusion, I will use the term spatial econometrics throughout the rest of the book. [8]

2.2.   Spatial Effects.

Spatial effects are the essential reason for the existence of a separate field of spatial econometrics.   As pointed out earlier, spatial dependence and spatial heterogeneity are the two aspects of data and models in regional science which merit particular attention from a methodological standpoint.   I now briefly discuss each in more specific terms.

2.2.1. Spatial Dependence.

In most empirical exercises in applied regional science, data are obtained for observations which are ordered in space or in space and time.   In these situations, the observations can be characterized by their absolute location, using a coordinate system, or by their relative location, based on a particular distance metric.   In other words, the data are organized by spatial units of observation, in the most general sense. [9]

Familiar examples of this empirical situation are the use of data on population, employment, and other economic activity collected for administrative units such as states, provinces, counties, or census tracts, located in geographic space.   Other typical instances of this occur when observations are organized by grid cells in an artificially constructed coordinate system, such as a digitized base map of a geographic information system.   In a more general sense, any situation where data are structured subject to a measure of location or distance, in any space, may be considered.   For example, measures derived for clusters of industrial activities characterized by a sectoral profile (i.e., a point in multi−dimensional profile space), or information on nations or interests groups identified by their positions in policy space would satisfy this requirement.

This type of data is often used by local and regional planning agencies which need to provide employment, population and housing projections and socio−economic impact analyses.   For example, studies of regional industrial structure may use census of manufacturers data on counties in a state, analyses of housing markets may be based on data for census tracts from the decennial census, and small area econometric models may be estimated using data on a number of contiguous counties, in order to alleviate the paucity of relevant time series of reasonable length.

As pointed out in the previous section, one of the main methodological problems present in these instances follows from the existence of spatial dependence. In general terms, spatial dependence can be considered to be the existence of a functional relationship between what happens at one point in space and what happens elsewhere.   Two broad classes of conditions would lead to this.   The first is a byproduct of measurement errors for observations in contiguous spatial units.   The second is more fundamental to regional science and human geography, and follows from the existence of a variety of spatial interaction phenomena.   First, I consider a simple example to illustrate how measurement error can easily result in spatial dependence.

In many situations encountered in practice, data is collected only at an aggregate scale. Therefore, there may be little correspondence between the spatial scope of the phenomenon under study and the delineation of the spatial units of observation. As a consequence, measurement errors are likely. Moreover, they will tend to spill over across the boundaries of the spatial units. As a result, the errors for one observation, say in spatial unit i, are likely to be related to the errors in a neighboring unit j. This spatial spill−over in measurement errors is one obvious cause for the presence of spatial dependence.

In a regression context, this can easily lead to non−spherical disturbance terms and errors in variables problems, as illustrated by the hypothetical situation shown in Figure 2.1. Here, the true spatial scales of the variables under consideration are the areas A, B and C, while observations are aggregated at levels 1 and 2. As a result, the observed variable Y_1 will be an aggregate of Y_A and part of Y_B, and the observed variable Y_2 will be an aggregate of Y_C and the remainder of Y_B:

$$Y_1 = Y_A + \lambda.Y_B$$

$$Y_2 = Y_C + (1-\lambda).Y_B$$

This aggregation is likely to suffer from errors in the assessment of the weighting parameter λ, which is present in Y_1 as well as in Y_2. As a result, these measurement errors will generate a pattern which exhibits spatial dependence.

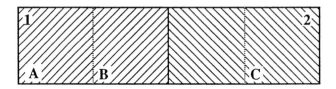

Figure 2.1.   Spatial Dependence and Aggregation

The second factor which may cause spatial dependence is more fundamental, and follows from the importance of space as an element in structuring explanations of human behavior. The essence of regional science and human geography is that location and distance matter, and result in a variety of interdependencies in space−time. Spatial interaction theories, diffusion processes, and spatial hierarchies yield formal frameworks to structure the dependence between phenomena at different locations in space. As a result, what is observed at one point is determined (in part) by what happens elsewhere in the system. This can formally be expressed in a spatial process, as illustrated below:

$$y_i = f(y_1, y_2, ..., y_N)$$

Here, every observation on a variable y at $i \in S$ (with S as the set containing all spatial units of observation) is related formally through the function f to the magnitudes for the variable in other spatial units in the system.

This simple expression in itself is not very useful in an empirical situation, since it would result in an unidentifiable system, with many more parameters (potentially N^2-N) than observations (N). By imposing a structure on the functional relationships embedded in f, i.e., a particular form for the spatial process, a limited number of characteristics of the spatial dependence may be estimated and tested empirically. This is the basis for the approaches towards model specification and estimation in spatial econometrics that are discussed in the next chapters.

Before dealing with the second type of spatial effect, spatial heterogeneity, it is important to reiterate the distinction between two different approaches towards the modeling of spatial dependence. One starts from theory, and posits a structure for spatial dependence a priori. This structure is incorporated in a formal model specification which is then used in the statistical analysis. The other approach departs from the data, and one attempts to infer an appropriate form for the dependence from a number of indicators, such as autocorrelation and cross—correlation statistics. This second approach, fundamental to exploratory data analysis and spatial time series analysis, will not be considered in detail in this book. [10]

2.2.2. Spatial Heterogeneity

In the literature of regional science and economic geography, there is ample evidence for the lack of uniformity of the effects of space. Several factors, such as central place hierarchies, the existence of leading and lagging regions, vintage effects in urban growth, etc., would argue for modeling strategies that take into account the particular features of each location (or spatial unit). In econometric work, this can be carried out by explicitly considering varying parameters, random coefficients, or various forms of structural change, such as switching regressions.

In addition to this lack of structural stability of the various phenomena over space, the spatial units of observation themselves are far from homogeneous. For example, census tracts have different area and shape, urban places have unequal populations or income levels, and regions have various degrees of technological development. To the extent that these aspects of heterogeneity are reflected in measurement errors (missing variables, functional misspecification) they may result in heteroskedasticity.

These various aspects of spatial heterogeneity are easiest illustrated in a regression context where cross sectional data are combined with time series data, as in the following general expression:

$$y_{it} = f_{it} (x_{it}, \beta_{it}, \epsilon_{it}).$$

Here, the index i refers to a spatial unit of observation, and t to the time period. The f_{it} is a time—space specific functional relationship which explains the value of the dependent variable y_{it} (or a vector of dependent variables) in terms of a vector of independent variables x_{it}, a vector of parameters β_{it}, and an error term ϵ_{it}. Of course, this formulation is not operational, since there are more parameters than observations. In order to carry out effective estimation and inference, and to ensure the identifiability of the model, a number of constraints need to be imposed on this general expression. In other words, it will be necessary to trade off the locational specificity in the model for identifiability of the parameters and functional forms, within the constraints imposed by data availability. In practice, the lack of sufficient data will often be severe, so that it is imperative that the simplification of the model is carried out judiciously. This problem is even more complex when the spatial heterogeneity occurs in combination with spatial dependence, as alluded to earlier.

The particular contribution of spatial econometrics to this problem consists of applying insights from regional science theory on spatial structure and spatial interaction as the basis for the various constraints and reparameterizations. As in the case of spatial dependence, this theoretical insight forms the point of departure for the selection and implementation of a model specification that provides structure to an otherwise intractable empirical problem.

NOTES ON CHAPTER 2

[1]   See, Hordijk and Paelinck (1976, pp. 175–9), Paelinck and Klaassen (1979, p. vii, pp. 5–11), Paelinck (1982, pp. 2–3).

[2]   For example, a recent book on *Urban Econometrics*, by Kau, Lee and Sirmans (1986) deals with the econometric analysis of urban economic models, but largely ignores spatial effects which may affect this analysis.

[3]   It is also very relevant in the physical sciences, such as plant ecology, geology, epidemiology, etc.   However, since these fields are not central to applied regional science, they will not be further considered.   Interested readers are referred to Cliff and Ord (1981), Ripley (1981), Gaile and Wilmott (1984), and Upton and Fingleton (1985) for extensive references.

[4]   In the literature the terms spatial structure, spatial pattern and spatial process are often ill–defined and used interchangeably.   Here, I follow the distinction made by Haining (1986a, p. 59–60) between spatial flow, spatial pattern and spatial structure. Spatial flow is taken to be the "physical transfer of commodities, people, information," etc. Spatial structure refers to the "background geography" and is rather fixed.   Spatial pattern "relates to more volatile or changeable levels of spatial regularity that may be imposed on the more permanent structure."   Spatial processes are phenomena which relate the three elements.   See also Bennett, Haining and Wilson (1985), for an overview.

[5]   For a recent overview, see Hsiao (1986).   An exception here is the work by Scott and Holt (1982), and King and Evans (1985, 1986) on cross–sectional dependencies resulting from block structures in survey data.   However, a spatial interpretation of these dependencies is not given.

[6]   A similar notion is spatial contextual variation, as proposed by Casetti (1972, 1986) as the motivation for a spatial expansion method.   However, this method is primarily concerned with a particular form of parameter variation over space.   The notion of spatial heterogeneity used here is more encompassing, and includes other forms of variability, such as heteroskedasticity and functional change.

[7]   In spatial point pattern analysis, this issue is known as the problem of real contagion versus apparent contagion, and has been studied extensively.   See, e.g., Getis and Boots (1978), Diggle (1983), Upton and Fingleton (1985).

[8]   See the more detailed comments in Haining (1986b) and Anselin (1986b).

[9]   The organization of observational units in time is not specific to spatial econometrics and will not be further considered here.   This issue can be fruitfully dealt with using standard econometric techniques, whereas the spatial organization has many complicating features that require special methodological tools.

[10]   An extensive treatment is given in Bennett (1979), to which the interested reader is referred.

CHAPTER 3

THE FORMAL EXPRESSION OF SPATIAL EFFECTS

To facilitate the reading of the later chapters, I present an initial discussion, in fairly general terms, of the notion of connectedness in space, and the main operational tools by which spatial effects are encompassed in econometric work: the spatial weight matrices and spatial lag operators. I also briefly discuss some complicating factors related to the notion of space implied by the various techniques.

3.1. The Formal Expression of Connectivity in Space

One of the crucial operational issues in spatial econometrics is the problem of formally expressing the way in which the structure of spatial dependence is to be incorporated in a model. In contrast to the situation in time series analysis, where the notion of a lagged variable is fairly unambiguous, in spatial analysis matters are complicated considerably. Not only is the definition of a spatial lag in two dimensions largely arbitrary, but the extension of this concept to higher order lags is not without problems. Moreover, the treatment of this issue in the literature is far from uniform, and a variety of terms and approaches abound.

In the remainder of this section, I survey various concepts and definitions which are relevant in formally expressing the notion of spatial dependence. Although the literature on this issue alone is quite extensive, I will limit the discussion to selected aspects that are particularly important in terms of the model specification and estimation issues treated in later chapters. [1] Throughout, it is assumed that observations are organized in spatial units, which may be points on a regular or irregular lattice, or regions on a map.

3.1.1. Neighbors in Space

The very notion of spatial dependence implies the need to determine which other units in the spatial system have an influence on the particular unit under consideration. Formally, this is expressed in the topological notions of neighborhood and nearest neighbor.

Consider a system S of N spatial units, labeled i=1,2,...N, and a variable x observed for each of these spatial units. In the literature on lattice processes and random field models, a set of neighbors for a spatial unit i is defined as the collection of those units j for which x_i is contained in the functional form of the conditional probability of x_i, conditional upon x at all other locations. [2] Formally, this definition would yield the set of neighbors for i as J, for which:

$$P[x_i \mid x] = P[x_i \mid x_J]$$

where x_J is the vector of observations for $x_j \ \forall \ j \in J$, and x is the vector containing all x values in the system. Alternatively, and less strictly, the set of neighbors j for i can be taken as

$$\{ j \mid P[x_i] \neq P[x_i \mid x_j] \ \}$$

or, as those locations for which the conditional marginal probability for x_i is not equal to the unconditional marginal probability. Note that neither of these definitions includes information about the relative location of the two spatial units, but only pertains to the influence via conditional probabilities.

In order to introduce a *spatial* aspect to these definitions of neighbors, which also makes the link with the notions of spatial stochastic processes discussed in Chapter 5, I suggest the following alternative working definition:

$$\{ j \mid P[x_i] \neq P[x_i \mid x_j] \ \text{and} \ d_{ij} < \epsilon_i \ \}$$

where d_{ij} is a measure of the distance between i and j in a properly structured space, and ϵ_i is a critical cut−off point for each spatial unit i, and possibly the same for all spatial units. The distance metric underlying d_{ij} is in the most general sense and can pertain to a Euclidean, Manhattan Block or general Minskowski distance.

This alternative concept of neighbor introduces an additional structure in the spatial data set, by combining a notion of statistical dependence (relating magnitudes) with a notion of space (distance and relative location). Although this definition does not preclude spatial units j that do not meet the distance criterion from exerting an influence on the conditional probability for x_i, they are not considered to be nearest neighbors. However, they can be included as higher order neighbors, which implies that the influence of j on i works via other spatial units. The definition based on conditional probabilities only does not allow for this distinction between first order and higher order neighbors.

The resulting set of neighbors for each spatial unit can be represented in a graph or network structure and associated connectivity matrix, which I discuss next.

3.1.2. Spatial Contiguity Matrices

The original m.. ures for spatial dependence, or, more precisely spatial autocorrelation, advanced by Moran (1948) and Geary (1954) were based on the notion of binary contiguity between spatial units. According to this notion, the underlying structure of neighbors is expressed by 0−1 values. If two spatial units have a common border of non−zero length they are considered to be contiguous, and a value of 1 is assigned.

This definition of contiguity obviously assumes the existence of a map, from which the boundaries can be discerned. For an irregular arrangement of spatial areal units this is fairly straightfoward. However, when the spatial units refer to a regular grid, or to a collection of irregularly arranged points, the determination of contiguity is not unique.

For example, consider the regular grid and associated centroids illustrated in Figure 3.1. A *common border* between cell a and surrounding cells can be considered in a number of different ways. For example, it can be taken as a common edge, with the cells labeled b as contiguous. Alternatively, a common vertex could be considered, with the cells labeled c as contiguous, or a combination of both notions could be used. In analogy with the game of chess, these situations have been called the rook case, the bishop case and the queen case. [3]

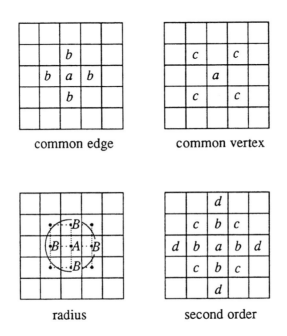

Figure 3.1. Contiguity on a Regular Lattice

When the spatial units consist of points, such as cities in an urban hierarchy, that are regularly or irregularly spaced over the system, the meaning of contiguity can be derived from the notion of shortest path on a network (or graph) formed by connecting the points, such as the network formed by the dashed lines in Figure 3.1. Nodes on the network are considered as neighbors if they are within a given maximum (shortest path) distance of each other. For example, in Figure 3.1, the nodes labeled B fall within radius d from centroid A, and can be considered as contiguous for this particular critical distance. [4]

Alternatively, the boundaries generated by various spatial tessellations could be considered to determine contiguity. These tessellations consist of an areal division of the spatial system into polygons or cells that are related in a systematic manner to the location of the points. [5] Consequently, the original representation of spatial units by points is replaced by a map of polygons (or regions), for which contiguity can be determined in the usual fashion.

In a similar manner, several orders of contiguity can be considered. This is achieved in a recursive way, by defining k−th order contiguity when spatial units are first order contiguous to a (k−1)−th order contiguous spatial unit, and not already contiguous of a smaller order. In a square grid system, this corresponds to a series of concentric bands around the spatial unit under consideration. For example, in Figure 3.1, the cells labeled c and d are second order contiguous to a according to the rook criterion, since they are first order contiguous to b.

Similar to the approach taken for a graph or network, the resulting spatial structure is formally expressed in a contiguity or connectivity matrix. In this matrix, each spatial unit is represented both as a row and as a column. In each row, the nonzero column elements correspond to contiguous spatial units. For example, for the nine cells in the center of Figure 3.1, the corresponding 9 by 9 matrix (with the cells numbered from left to right and top to bottom) is given in Table 3.1, using the rook definition of contiguity. By convention, a cell is not contiguous to itself, which results in zero diagonal elements.

Table 3.1.   Binary Contiguity Matrix for Gridded Data

0	1	0	1	0	0	0	0	0
1	0	1	0	1	0	0	0	0
0	1	0	0	0	1	0	0	0
1	0	0	0	1	0	1	0	0
0	1	0	1	0	1	0	1	0
0	0	1	0	1	0	0	0	1
0	0	0	1	0	0	0	1	0
0	0	0	0	1	0	1	0	1
0	0	0	0	0	1	0	1	0

Clearly, the great variety of ways in which binary contiguity can be formalized is not a desirable feature. Moreover, simple contiguity only provides a limited representation of the extent of spatial interaction that can be expressed in a model. In addition, it is not sensitive to a number of topological transformations, in the sense that the same contiguity matrix can represent many different arrangements of the spatial units. [6]

3.1.3. General Spatial Weight Matrices

The simple concept of binary contiguity was extended by Cliff and Ord (1973, 1981) to include a general measure of the potential interaction between two spatial units. This is expressed in a spatial weight matrix W, also referred to as a Cliff–Ord weight matrix. The determination of the proper specification for the elements of this matrix, w_{ij}, is one of the more difficult and controversial methodological issues in spatial econometrics.

The original suggestion by Cliff and Ord consists of using a combination of distance measures (inverse distance, or negative exponentials of distance) and the relative length of the common border between two spatial units, in the sense of the share in the total border length that is occupied by the other unit under consideration. The resulting weights will therefore be asymmetric, unless both spatial units have the same total boundary length. Formally:

$$w_{ij} = [d_{ij}]^{-a}.[\beta_{ij}]^b$$

with d_{ij} as the distance between spatial unit i and j, β_{ij} as the proportion of the interior boundary of unit i which is in contact with unit j, and a and b as parameters.

In a similar vein, Dacey (1968) suggested weights that also take into account the relative area of the spatial units:

$$w_{ij} = d_{ij}.\alpha_i.\beta_{ij}$$

with d_{ij} as a binary contiguity factor, α_i as the share of unit i in the total area of all spatial units in the system, and β_{ij} as the boundary measure used above.

Both these weights are closely linked to the physical features of spatial units on a map. As with the binary contiguity measures, they are less useful when the spatial units consist of points, since then the notions of boundary length and area are largely artificial, and determined by a particular tessellation algorithm. They are also less meaningful when the spatial interaction phenomenon under consideration is determined by factors such as purely economic variables, which may have little to do with the spatial configuration of boundaries on a physical map.

Consequently, several authors have suggested the use of weights with a more direct relation to the particular phenomenon under study. For example, Bodson and Peeters (1975) introduced a general accessibility weight (calibrated between 0 and 1), which combines in a logistic function the influence of several channels of communication between regions, such as roads, railways, and other communication links. Formally:

$$w_{ij} = \Sigma_j \ k_j.\{a \ /[1 \ + \ b.exp(-c_j d_{ij})]\}$$

in which k_j shows the relative importance of the means of communication j. The sum is over the j means of communication, which separate the spatial units by a distance of d_{ij}. The a, b and c_j are parameters, which need to be estimated.

In most applications in regional science, the weight matrix is based on some combination of distance relationships and simple contiguity. The distances involved are in the most general sense, and can be based on travel time, *general distance* (in the sense of Isard, 1969), or derived from a multidimensional scaling analysis

(Gatrell, 1979). In most of the sociological applications of spatial analysis, the weight matrix is determined by concepts from social network theory. [7]

An important problem results from the incorporation of parameters in the weights. Typically, these weights are taken to be exogenous and the parameter values are determined a priori, or in a step separate from the rest of the spatial analysis. [8] This creates problems for the estimation and interpretation of the results. In particular, it could potentially lead to the inference of *spurious* relationships, since the validity of estimates is pre−conditioned by the extent to which the spatial structure is correctly reflected in the weights. More importantly, it could result in a circular reasoning, in that the spatial structure, which the analyst may wish to discover in the data, has to be assumed known before the data analysis is carried out. I will return to this important issue in later chapters.

Overall, there is no agreement as to which type of weight matrix should be used in spatial econometric analysis. Arora and Brown (1977) and Hordijk (1979) suggest the use of *neutral* weights when dealing with spatial models for error terms. They identify neutral with binary contiguity. However, Hordijk (1979) also argues for the use of a *general* weight matrix in the functional specification of spatial econometric models (i.e., not for the error terms), the elements of which should be specified a priori.

In Anselin (1980, 1984a), I argue that the structure of spatial dependence incorporated in the spatial weight matrix should be chosen judiciously, and related to general concepts from spatial interaction theory, such as the notions of accessibility and potential. In line with a model−driven approach to spatial econometrics, the weight matrix should bear a direct relation to a theoretical conceptualization of the structure of dependence, rather than reflecting an ad hoc description of spatial pattern.

When the weight matrix is used in a hypothesis test, this requirement is less stringent. Since the null hypothesis is one of spatial independence, the weight matrix should be related to the relevant alternative hypothesis of spatial dependence, in order to maximize the power of the test. However, even with an improperly specified weight matrix, a conservative interpretation of a rejection of the null will only imply a lack of independence, and not a particular type of dependence. Although the power of the test will be affected, the potential for spurious conclusions is not as great as in the functional specification of a model.

Finally, it should be noted that the often cited argument for the use of simple binary contiguity weights, as a parallel to time series analysis, does not really hold. Indeed, from a strict statistical point of view, the necessary assumptions and the implications for the allowable spatial structure are very restrictive. As outlined in Hooper and Hewings (1981), the formal requirements are only satisfied for a few regular lattice structures that have little relevance for empirical analysis in applied regional science. This issue is discussed in more detail in Section 5.1, where spatial stochastic processes are considered from a formal standpoint.

3.1.4. Spatial Lag Operators

The ultimate objective of the use of a spatial weight matrix in the specification of spatial econometric models is to relate a variable at one point in space to the observations for that variable in other spatial units in the system. In a time series context, this is achieved by using a lag operator, which shifts the variable by one or more periods in time (e.g., Box and Jenkins 1976, Dhrymes, 1981). For example:

$$y_{t-k} = L^k y$$

shows the variable y shifted k periods back from t, as a k−th power of the lag operator L.

In space, matters are not this straightforward, due to the many directions in which the shift can take place. As an illustration, consider the regular lattice structure in Figure 3.2. The variable x, observed at location i,j, can be shifted in the following ways, using the simple contiguity criteria:

− using a rook criterion of contiguity, to:

$$x_{i-1,j}; \; x_{i,j-1}; \; x_{i+1,j}; \; x_{i,j+1};$$

− using a bishop criterion of contiguity, to:

$$x_{i-1,j-1}; \; x_{i+1,j-1}; \; x_{i+1,j+1}; \; x_{i-1,j+1};$$

For a queen type of contiguity, the number of possible locations increases to a total of eight.

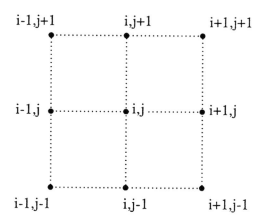

Figure 3.2   Spatial Lags on a Regular Lattice

In most applied situations, there are no strong a priori motivations to guide the choice of the relevant direction of dependence. This problem is compounded when the spatial arrangement of observations is irregular, since then an infinite number of directional shifts becomes possible. Clearly, the number of parameters associated with all shifted positions quickly would become unwieldy and preclude any meaningful analysis. Moreover, unless the data set is very large and structured in a regular way, the remaining degrees of freedom would be insufficient to allow an efficient estimation of these parameters.

This problem is resolved by considering a weighted sum of all values belonging to a given contiguity class, rather than taking each of them individually. The terms of the sum are obtained by multiplying the observations in question times the associated weight from the spatial weight matrix. Formally:

$$L^s x_i = \Sigma_j \ w_{ij}.x_j \quad \forall \ j \in J_i$$

where L^s is the lag operator associated with contiguity class s, j is the index of the observations belonging to the contiguity class s for i, and w_{ij} are the spatial weights. In matrix notation, with all observations in the system contained in a vector x, this becomes, for contiguity class s:

$$L^s x = W_s x.$$

Clearly, the resulting notion of a spatially lagged variable is not the same as in time series analysis, but instead is similar to the concept of a distributed lag. In this regard, it is important to note that the weights used in the construction of the lagged variables are taken as given, just as a particular time path can be imposed in the estimation of a distributed lag. The joint determination of the weights and measures of statistical association, such as correlation or regression coefficients becomes a non−linear problem. However, by fixing the weights a priori, this is reduced to a more manageable linear problem, at the risk of imposing a potentially wrong structure.

A less restrictive spatially lagged variable can be constructed from the notion of potential or accessibility, as:

$$f_i = \Sigma_j \ q(d_{ij},\theta).x_j$$

where f_i is the potential at i, and q is a function of distance d_{ij} between i and the other spatial units j, parameterized in terms of a vector of coefficients θ. [9] Since the resulting expression is non−linear, estimation and hypothesis testing will be more complex.

The actual weight matrix W used in the spatial lag is often standardized such that the row elements sum to one. Although there is no mathematical or statistical requirement for this, in many instances it facilitates the interpretation of the model coefficients. On the other hand, the standardized matrix is usually not symmetric, which has implications for the numerical complexity of estimation and testing procedures.

The standardization of the weight matrix should not be carried out automatically. In fact, when the weights are based on an inverse distance function or a similar concept of distance decay, which has a meaningful economic interpretation, scaling the rows so that the weights sum to one may result in a loss of that interpretation.

The interpretation of the spatial interaction implied by a weight matrix is not always economically meaningful. For example, when a simple contiguity forms the basis for the weights, a scaling implies a simple average of the values for the contiguous units. In other words, if there are M contiguous units, the lagged variable is $1/M$ times the sum of the relevant x_j. In spatial interaction terms, this implies a competition between neighboring units: the fewer neighbors, the stronger their individual influence on the central unit will be. Clearly, this particular view of spatial interaction does not always make sense in an economic context.

3.1.5. Circularity and Redundancy in Spatial Lag Operators

Similar to the situation with a time lag in time series analysis, the concept of a spatial lag can be extended to higher orders. However, in a spatial context the higher order refers to a different contiguity class, rather than to a longer lag. Moreover, this concept only has a precise meaning for a binary contiguity matrix.

As pointed out before, the contiguity matrix can be interpreted as a representation of the structure of linkages in a network. For example, a first order contiguity matrix shows all paths of length one that exist in the graph. In network terminology, a higher order contiguity would correspond to a path of a longer length. This can be expressed in formal terms by powering the first order contiguity matrix. The resulting connectivity matrices show the various linkages between nodes with a path length equal to the power. More precisely, the off-diagonal elements of the powered matrix show the total number of connecting paths between each pair of units in the system. Since the existence of a path is equivalent to the notion of contiguity, this provides an easy way to construct higher order spatial weight matrices.

Blommestein (1985) has pointed out that this procedure can easily result in circular routes or redundant paths. This means that the powers of the contiguity matrix include paths that are already partially contained in a contiguity matrix of a lower order. Consequently, an uncritical interpretation of these linkages would lead to double—counting. This redundancy has implications for estimation and inference. Therefore, the powered matrices should be corrected to eliminate the circular paths before they are used to construct spatially lagged variables.

To illustrate this problem, consider the matrices in Table 3.2 and Table 3.3, which correspond to the second and third power of the first order contiguity matrix for a 3 by 3 regular lattice (rook case, as in Table 3.1). The non—zero elements in a row point to the existence of a link of length two and three respectively. When possible circular paths are ignored, these elements would be assigned to the corresponding contiguity class.

Table 3.2   Second Power of a Binary Contiguity Matrix

0	0	1	0	2	0	1	0	0
0	0	0	2	0	2	0	1	0
1	0	0	0	2	0	0	0	1
0	2	0	0	0	1	0	2	0
2	0	2	0	0	0	2	0	2
0	2	0	1	0	0	0	2	0
1	0	0	0	2	0	0	0	1
0	1	0	2	0	2	0	0	0
0	0	1	0	2	0	1	0	0

Table 3.3.   Third Power of a Binary Contiguity Matrix

0	5	0	5	0	3	0	3	0
5	0	5	0	8	0	3	0	3
0	5	0	3	0	5	0	3	0
5	0	3	0	8	0	5	0	3
0	8	0	8	0	8	0	8	0
3	0	5	0	8	0	3	0	5
0	3	0	5	0	3	0	5	0
3	0	3	0	8	0	5	0	5
0	3	0	3	0	5	0	5	0

The values in the rows give the number of alternative paths along which a linkage between row and column cell can be obtained. For example, in the first row of Table 3.2, the fifth element (2) shows that a two−step linkage between cell 1 and cell 5 can be obtained along two different paths: 5−2−1 and 5−4−1. However, this particular information is not taken into account for the purposes of constructing a spatial contiguity matrix, since the non−zero values in the powered matrix are set to one. Also, by convention, all diagonal elements are ignored and set to zero.

The redundancy becomes clear in the third power matrix, when its first row is compared to the first row of the first order contiguity matrix: elements two and four are common to both. In a strict definition of contiguity, these linkages cannot be considered as third order contiguity, since the corresponding cells are already contiguous of a lower (first) order. Although this may seem counterintuitive, it should be clear that there are paths in the network so that node two can reach node one directly, as well as via another node in three steps (e.g., 2−5−4−1). For the specification and estimation of econometric models that include spatial dependence, this is an undesirable characteristic. [10]

In contrast to the binary contiguity case, a higher order lag based on general weights does not have a clear interpretation in terms of a different spatial pattern of linkages. It this case, the weight matrix is more fruitfully interpreted as a reduced form, which summarizes a host of spatial (or space—time) interactions in the system, based on insights from regional science theory. When the situation requires the inclusion of several patterns of interaction, they can be entered in their own right, as separate spatially lagged variables, as in Brandsma and Ketellapper (1979a). This will avoid creating the impression of a similarity with higher order lags in time series analysis, which should be rejected as artificial and often misleading. [11]

3.2. Problems with the Concept of Space in the Formal Expression of Spatial Effects

In this section, I will briefly discuss some issues which follow from the lack of a uniform organization of spatial data. In contrast to the situation in time series analysis, where the choice of a time axis is fairly unambiguous, the variety of spatial weight matrices and levels of spatial aggregation create specific methodological problems for spatial econometrics.

One important issue is the modifiable areal unit problem, which seems to call into question the generality of statistical measures of spatial effects. This is addressed first. Another issue follows from the lack of uniform criteria for the choice of a spatial weight matrix. Since the properties of estimators and tests will be determined in part by the form of spatial dependence incorporated in these weights, it is important that the matrices can be characterized in formal terms.

3.2.1. The Modifiable Areal Unit Problem

A substantial part of the econometric analysis in applied regional science is based on data collected for spatial units with irregular and arbitrary boundaries. Nonetheless, the interpretation of the various models and the implications for policy are often made with respect to a general notion of *space*. This would imply that there is some unique and identifiable spatial structure, with clear statistical properties, independent from the way in which the data is organized in spatial units. Unfortunately, matters are not this straightforward.

The modifiable areal unit problem pertains to the fact that statistical measures for cross—sectional data are sensitive to the way in which the spatial units are organized. Specifically, the level of aggregation and the spatial arrangement in *zones* (i.e., combinations of contiguous units) affects the magnitude of various measures of association, such as spatial autocorrelation coefficients and parameters in a regression model.

This problem is an old one, and is variously referred to as the micro—macro aggregation problem in econometrics and the ecological fallacy problem in sociology.

Its ramifications for spatial analysis were brought to the attention of geographers and regional scientists in two recent papers by Openshaw and Taylor (1979, 1981). Under the provocative title of "a million or so correlation coefficients" they illustrated how a spatial autocorrelation measure (a Moran I coefficient for first order contiguity) varied with different areal aggregations for data simulated over the spatial structure of Iowa counties. [12] This line of reasoning can easily be extended to the measure of spatial dependence itself, which can be shown to vary with a different choice of spatial weight matrix. Since the resulting statistical indications can range from strong association to absence of association, the modifiable areal unit problem would seem to seriously call into question the validity of such spatial analysis. [13]

However, upon closer examination, the modifiable areal unit problem can be considered as a combination of two familiar problems in econometrics: aggregation and identification.

The first aspect of the problem pertains to the aggregation of spatial units. As is well known, this is only meaningful if the underlying phenomenon is homogeneous across the units of observation. If this is not the case, the inherent heterogeneity and structural instability should be accounted for in the various aggregation schemes. In other words, unless there is a homogeneous spatial process underlying the data, any aggregation will tend to be misleading. [14] Consequently, this aspect of the modifiable areal unit problem should be considered as a specification issue, related to the form of spatial heterogeneity, and not solely as an issue determined by the spatial organization of the data.

The second aspect pertains to the proper identification of the structure of spatial dependence. As pointed out in the previous sections, an analysis of spatial association is typically carried out by relating a variable to its spatially lagged counterpart. The latter is constructed as a linear combination of the observations in the system. Typically, the association is indicated by a correlation or regression coefficient. For example, a variable y would be related to ρWy, where ρ is a spatial autoregressive coefficient, and W is a spatial weight matrix. An implication from the areal unit literature is that a different choice of W will result in a different ρ, and that therefore the measure of spatial association is indeterminate.

From an econometric standpoint, the problem can be viewed as an identification problem, since there is insufficient information in the data to allow for the full specification of the simultaneous interaction over space. In this sense, a formulation of linear spatial association can be considered as a special case of a system of simultaneous linear equations, with one observation for each equation:

$$y_i = \Sigma_j \ \beta_{ij} y_j, \ \forall \ i \ \text{in the system.}$$

Obviously, constraints need to be imposed for at least some of the model parameters to be identifiable. Such constraints can be pertain to the model coefficients or to the error structure. [15] As outlined above, the usual approach in spatial analysis is to introduce a spatially lagged variable and thereby to reduce the empirical problem to that of estimating one parameter, ρ, in:

$$y_i = \rho \; \Sigma_j \; w_{ij}y_j.$$

Clearly, a choice of different weights w_{ij} is likely to result in a different estimate for ρ.

This seeming indeterminacy of ρ is mostly a problem in an exploratory data analysis, since there is insufficient structure in the data as such to derive the proper spatial model. In the model—driven approach which is taken here, a priori (theoretical) reasons dictate the particular form for the identification constraints, similar to the approach taken for systems of simultaneous equations. Competing specifications can subsequently be compared by means of model specification tests and model selection procedures.

Given the importance of the choice of the proper constraints (weights) for the interpretation of spatial models, some inherent characteristics of weight matrices are considered next.

3.2.2. Properties of Spatial Weight Matrices

The intricate relationship between measures of spatial association and the choice of a connectivity matrix has several implications for the performance of estimators and test statistics. From a purely methodological standpoint, it is important to relate the properties of these various techniques to the degree of spatial connectedness or inherent spatial dependence which may be present in the data. This is particularly relevant for the design of Monte Carlo simulation experiments to assess the performance of estimators and tests in situations which mimick realistic empirical contexts.

In the case of regular square or rectangular lattice stuctures in the plane, of finite or infinite dimensions, the geometry of the problem allows for many simplifying results. Consequently, the structure of the spatial connectivity is well understood and can be summarized in elegant mathematical terms. [16] Unfortunately, these results are of little use for the irregular spatial configurations more often encountered in applied work. For these, no definite results are yet available.

This issue has primarily been dealt with in the geographical literature, where the analysis is limited to the binary contiguity matrix. The starting point is the equivalence of this matrix to the connectivity matrix of a network. The structure of this network is analyzed by means of concepts from graph theory. The factors which are most often used to characterize the inherent spatial structure are the size of the network (e.g., the number of observations), the average number of linkages per spatial unit, indicators of shape, and, most importantly, the principal eigenvalue of the contiguity matrix. The latter has been found by a number of authors to adequately summarize several features of the spatial structure of the network. [17]

The more interesting case of a general spatial weight matrix has yet to be studied in this respect. A fruitful avenue for research would be to exploit a certain analogy to the summary measures for interconnectedness in input—output models. Indeed, the technical coefficient matrix of a square input—output model shows many formal similarities to a spatial weight matrix. Consequently, measures of overall interconnectedness that are based on the direct coefficients (so—called technical measures) would seem to be equally applicable to summarize the overall connectivity reflected in a spatial weight matrix. Examples of these would be the percentage nonzero coefficients, row and column sums and means, the Yan and Ames interrelatedness index, and measures derived from information theory (Shannon's formula). [18] The application of these measures to spatial analysis merits further attention.

NOTES ON CHAPTER 3

[1] A further motivation for keeping this discussion brief is that the issue of spatial connectivity has received extensive attention in the texts of Cliff and Ord (1981) and Upton and Fingleton (1985). Therefore, I will concentrate on issues which have received less attention and on viewpoints that differ from the approach taken in these texts.

[2] For a more rigorous discussion, see Besag (1974), Bartlett (1978), and Haining (1979).

[3] For a more extensive discussion, see Upton and Fingleton (1985, Chapter 3).

[4] The shortest path on a network obtained from connecting all the points in the system is not the only way in which contiguity can be defined. Alternatively, the notion of Gabriel connectivity could be used. According to this concept, two points are considered contiguous if all other points in the system are outside a circle on which circumference the points are at opposite ends (the so-called least squares adjacency criterion). For details, see Matula and Sokal (1980).

[5] The best-known types of spatial tessellations are the Thiessen polygons (also called Dirichlet tessellation or Voronoi polygons), which are constructed from the perpendicular bisectors of the lines that connect the irregularly spaced points (for an overview, see Ripley 1981, pp. 38–44, Amrhein, Guevara and Griffith 1983, and Upton and Fingleton 1985, pp. 96–104). These polygons are often related to notions of spatial market areas, and can easily be used to obtain measures of contiguity. For example, in Figure 3.1, the square grids form Thiessen polygons for their centroids (points connected by the dashed lines). Other tessellations, such as the Delaunay triangularization (the dual of the Thiessen polygons), and various other mosaics are less directly relevant for applications in regional science.

[6] See Cliff and Ord (1973, p. 10), for an example.

[7] For overviews and applications, see, e.g., Pool and Kochen (1978), Burt (1980), White, Burton, and Dow (1981), Loftin and Ward (1983), and the papers by Doreian: Doreian (1974, 1980, 1981, 1982), and Doreian, Teuter and Wang (1984).

[8] An exception to the a priori choice of weights is the approach suggested by Kooijman (1976). There, an alternative test statistic for spatial autocorrelation is constructed from a constrained maximization process in function of the contiguity weights, somewhat similar to the technique of kriging, which is widely used in geological sciences. For an overview of kriging, see Matheron (1971), Clark (1979), Ripley (1981, pp. 44–50), and Nipper and Streit (1982). Hanham, Hohn and Bohland (1984) present an application to a demographic phenomenon. However, most applications of kriging are in the physical sciences, without much relevance for the types of problems considered in regional science. Therefore, this technique will not be further considered and the interested reader is referred to the sources listed above.

[9] A wide range of functional forms can be considered. Weibull (1976) outlines an axiomatic approach which restricts this choice to specific types of functions. See also Anselin (1980, Chapter 8), for a more extensive discussion.

[10] The precise effects on estimation are presented in detail in Blommestein (1985).

[11] This view is discussed at greater length in Anselin (1986b).

[12] The level of aggregation varied from 99 counties to 6 zones.

[13] See also the comments in Johnston (1984), and Openshaw (1984).

[14] This issue is a familiar one in the regional science literature, and known as the *regional homogeneity* problem. For recent overviews, see, e.g., Johnson (1975), Schulze (1977, 1987), and Lin (1985). It is further discussed in Section 10.1.

[15] For an overview of the identification problem in econometrics, see Fisher (1966) and Hausman (1983).

[16] For examples and more detailed discussion, see, e.g., Whittle (1954), Bartlett (1975), and Griffith (1987).

[17] Since an extensive discussion of these primarily geometric considerations are tangential to the main focus of the book, the interested reader is referred to the literature for more detailed treatments. The main ideas on this issue go back to Garrison and Marble (1964), Gould (1967), and Tinkler (1972). More recent discussions of the use of the principal eigenvalue and associated eigenvectors to characterize spatial configuration can be found in Boots (1982, 1984, 1985), Boots and Tinkler (1983), and Griffith (1984). A related issue of the information content in binary maps is discussed in Gatrell (1977).

[18] The Yan and Ames index consists of determining for each element of the contiguity matrix (which converts the weight matrix to a boolean matrix) how many powers are needed in order for it to become nonzero. An associated order matrix (which gives an indication of path length in the network) then forms the basis for the derivation of various summary indices, such as row, column, and overall means. A similar approach consists of using the weight matrix itself and considering how many powers are needed in order for an element of the matrix to become less than a pre—set small positive value. Details and a more extensive overview are given in Hamilton and Jensen (1983) and Szyrmer (1985).

CHAPTER 4

A TYPOLOGY OF SPATIAL ECONOMETRIC MODELS

A large number of model specifications for spatial processes have been suggested in the literature and empirically implemented. This great variety may seem unwieldy, and give the impression that every particular model necessitates its own methodological framework. Fortunately, some structure can be imposed, guided by the principle that econometric techniques can be applied in essentially the same manner to models grouped in terms of salient characteristics.

In order to organize the discussion and to set the stage for the treatment of estimation in Part II, in this chapter I suggest a number of lines along which spatial econometric models can be classified. As in the rest of the book, the classification is restricted to linear regression models. Consequently, some other familiar spatial multivariate models will not be considered. Examples of these are the multivariate spatial autocorrelation and correlation analysis by, e.g., Hubert, Golledge and Costanzo (1981), Hubert and Golledge (1982), Hubert (1985), Hubert, et al. (1985), and Wartenberg (1985), spatial analysis of variance by Griffith (1978), and spatial LISREL models, as discussed in Folmer and van der Knaap (1981), Folmer and Nijkamp (1984), and Folmer (1986).

First, I will focus on some general categories and discuss several criteria that may be used to guide the classification. Next, I present in more detail two taxonomies that I will use to structure the treatment in the rest of the book. One taxonomy pertains to spatial models designed for cross—sectional data, the other deals with space—time models.

4.1. A General Classification of Models

In Chapter 2, I argued that the main characteristic of spatial econometrics is the way in which spatial effects are taken into account. Of course, this pre-supposes that space has been formalized in one way or another. Typically, the use of a spatial weight matrix achieves that spatial models can be applied to many empirical contexts, provided that the spatial dependence is properly expressed in the weights and that the spatial heterogeneity is accounted for in the specification of the model.

In the earlier literature on lattice processes and random field models, the specification of the model was mostly in terms of a simple nearest—neighbor contiguity. Although these models exhibit some interesting formal mathematical properties, they have a limited scope for application in empirical regional science. Therefore, they will not be further considered. [1]

A fundamental distinction, with a significant impact on estimation and testing strategies, is the difference between a simultaneous and a conditional spatial process. The conditional model is based on a conditional probability specification, whereas the simultaneous model is expressed as a joint probability. Both models are most often presented in an autoregressive form, i.e., where the value of a variable at one point is related to its values in the rest of the spatial system.

The simultaneous model dates back to the work of Whittle (1954), in which random fields on a regular lattice were expressed in the form of a stochastic difference equation in the lattice coordinates:

$$\Sigma_s \ \Sigma_t \ \alpha_{m-s,n-t} \cdot y_{j+s,k+t} = \epsilon_{j,k}$$

with $j,k = ..., -1, 0, 1, ...,$ as spatial indices, y as the variable under consideration, the α as parameters, and the ϵ as independent normally distributed random variates. [2] In matrix notation, this is equivalent to a first−order spatial autoregressive model:

$$y = \rho.W.y + \epsilon.$$

with y and ϵ as vectors of variables and error terms, and W as the corresponding spatial weight matrix. The estimation of this model involves the specification of a joint probability distribution and necessitates nonlinear optimization.

The conditional model was originally suggested by Besag (1974) to obtain an alternative to the highly nonlinear estimation procedure of Whittle. In essence, it consists of a linear relationship between a conditional expectation of the dependent variable and its values in the rest of the system (S):

$$E\ [y_i \mid y_j \ \forall \ j \in S, \ j \neq i] = \rho.W.y.$$

The conditional structure of the model makes it possible for ordinary least squares to be used as an estimation techniques, provided that the data is coded in a particular manner. The coding essentially consists of eliminating observations so that the remaining dependent variables are *independent* (i.e., not contiguous). In a situation where few observations are available, this is clearly not a very practical approach. Also, the coding scheme is not unique and thus different estimates may be obtained for the same data set, depending on which observations are dropped.

More recently, Besag (1975, 1977), and Lele and Ord (1986) have developed a pseudo−maximum likelihood approach to the conditional model, which avoids the need for a coding scheme. However, the resulting estimation is fairly complex mathematically.

An interesting feature of the conditional approach is that it has been suggested as a way to avoid some of the problems associated with the specification of the spatial weight matrix (Haining, 1986a). Nevertheless, most modeling situations in empirical regional science are easier expressed in a simultaneous form. This form is also closer to the standard econometric approach. Therefore, given the model−driven approach taken in this book and the focus on econometric techniques, the emphasis will be almost exclusively on the simultaneous model. [3]

Analogous to the Box−Jenkins approach in time series analysis, spatial model specifications have been suggested that combine both autoregressive and moving average processes. [4] In this context, a spatial moving average model is of the form:

$$y = \epsilon + \rho W \epsilon$$

where y is a vector of variables, ϵ is a vector of independently distributed random terms, and W is the spatial weight matrix.

The moving average model introduces a different perspective on the spatial interaction in a spatial process. However, its consideration would add little to the complexity of estimation and model specification issues. Indeed, it turns out that most of the estimation and test results for the moving average model can be derived in a very similar manner to those for the autoregressive specification. [5] Similarly, most estimation aspects of STARIMA (space time autoregressive integrated moving average) models are not substantially different from the approaches used for the much simpler autoregressive form. [6] I will therefore limit the treatment in Part II to estimation and specification issues for spatial autoregressive models.

An additional consideration important for distinguishing between different model structures relates to data availability. For cross—sectional observations, there is insufficient information in the data to extract a fully simultaneous pattern of interaction, and the reparameterization of the model in terms of autoregressive structures is a necessity. However, when a combination of cross—sections and time series is available, the additional information from the time dimension may relax this requirement. Consequently, a much richer set of spatial modeling situations can be considered.

4.2.   A Taxonomy of Spatial Linear Regression Models for Cross—Section Data

In this section, I present a general specification, which forms a framework to organize various modeling situations of interest in spatial econometrics. The specification pertains to the situation where observations are available for a cross—section of spatial units, at one point in time. The approach taken consists of deriving specific models by imposing various constraints on the parameters of the general formulation. Consequently, in the treatment of estimation and specification which follows in Part II, many results for these models will be found in a simple manner as special cases of the derivation for the general model.

The point of departure is the following expression:

$$y = \rho W_1 y + X\beta + \epsilon \tag{4.1}$$

$$\epsilon = \lambda W_2 \epsilon + \mu$$

with $\mu \sim N(0,\Omega)$,

and the diagonal elements of the error covariance matrix Ω as:

$$\Omega_{ii} = h_i (z\alpha) \qquad\qquad h_i > 0.$$

In this specification, β is a K by 1 vector of parameters associated with exogenous (i.e., not lagged dependent) variables X (N by K matrix), ρ is the coefficient of the spatially lagged dependent variable, and λ is the coefficient in a spatial autoregressive structure for the disturbance ϵ. [7]

The disturbance μ is taken to be normally distributed with a general diagonal covariance matrix Ω. The diagonal elements allow for heteroskedasticity as a function of P+1 exogenous variables z, which include a constant term. The P parameters α are associated with the non−constant terms, such that, for $\alpha=0$, it follows that

$$h = \sigma^2,$$

(i.e., the classic homoskedastic situation).

The two N by N matrices W_1 and W_2 are standardized or unstandardized spatial weight matrices, respectively associated with a spatial autoregressive process in the dependent variable and in the disturbance term. [8] This allows for the two processes to be driven by a different spatial structure, e.g., as suggested in Hordijk (1979).

In all, the model has 3+K+P unknown parameters, in vectorform:

$$\theta = [\rho, \beta', \lambda, \sigma^2, \alpha']'. \tag{4.2}$$

Several familiar spatial model structures result when subvectors of the parameter vector (4.2) are set to zero. Specifically, the following situations correspond to the four traditional spatial autoregressive models discussed in the literature, as in, e.g., Hordijk (1979), Anselin (1980, 1988a), Bivand (1984):

− for: $\rho=0$, $\lambda=0$, $\alpha=0$ (P+2 constraints):

$$y = X\beta + \epsilon \tag{4.3}$$

i.e., the classical linear regression model, with no spatial effects.

− for: $\lambda=0$, $\alpha=0$ (P+1 constraints):

$$y = \rho W_1 y + X\beta + \epsilon \tag{4.4}$$

i.e., the mixed regressive−spatial autoregressive model, which includes common factor specifications (with WX included in the explanatory variables), as a special case.

− for: $\rho=0$, $\alpha=0$ (P+1 constraints):

$$y = X\beta + (I - \lambda W_2)^{-1}\mu \tag{4.5}$$

i.e., the linear regression model with a spatial autoregressive disturbance.

− for: $\alpha=0$ (P constraints):

$$y = \rho W_1 y + X\beta + (I - \lambda W_2)^{-1}\mu \tag{4.6}$$

i.e., the mixed—regressive—spatial autoregressive model with a spatial autoregressive disturbance.

Four more specifications are obtained by allowing heteroskedasticity of a specific form (i.e., a specific h(zα)) in the models (4.3)—(4.6).

This taxonomy focuses primarily on the specification of spatial dependence, although spatial heterogeneity can be incorporated in a straightforward way. Indeed, the most common forms of spatial heterogeneity are reflected in varying parameters or random coefficients. Since this will typically result in heteroskedasticity of some form, the heterogeneity can be encompassed by specifying the appropriate terms in the h(zα). Consequently, most situations of interest are taken into account in the general form (4.1).

4.3. A Taxonomy of Spatial Linear Regression Models for Space—Time Data

The introduction of the time dimension considerably increases the complexity of issues that can be taken into account in the specification of spatial econometric models. In more traditional econometric terms, this is the situation of pooled cross—section and time series data. This modeling framework is applied here to take into account patterns of cross—sectional dependence and heterogeneity.

As a point of departure, consider the following specification, which expresses a full range of potential space—time dependencies and forms of heterogeneity:

$$y_{it} = x_{it}\beta_{it} + \epsilon_{it} \tag{4.7}$$

where x_{it} is a row vector of observations for spatial unit i at time t, β_{it} is a vector of space—time specific parameters, and ϵ_{it} is an error term. The error term is characterized by the following conditions:

$$E\ [\epsilon_{it}] = 0 \tag{4.8}$$

$$E\ [\epsilon_{it} \cdot \epsilon_{js}] \neq 0. \tag{4.9}$$

Whereas expression (4.8) is a familiar condition, expression (4.9) provides the structure to encompasses a number of possible space—time dependencies and patterns of spatial heterogeneity.

For example, with i = j and t = s, the residual variance can be taken as constant (the standard case) or as varying, i.e., the heteroskedastic case. The variation can be across space (indexed by i), over time (indexed by t), or over space and time (indexed by i,t). Formally:

$$E\ [\epsilon_{it} \cdot \epsilon_{it}] = \sigma, \qquad \text{constant variance}$$

$$E\ [\epsilon_{it} \cdot \epsilon_{it}] = \sigma_i, \qquad \text{spatial heterogeneity}$$

$$E\ [\epsilon_{it} \cdot \epsilon_{it}] = \sigma_t, \qquad \text{time-wise heterogeneity}$$

$$E\ [\epsilon_{it} \cdot \epsilon_{it}] = \sigma_{it}, \qquad \text{space-time specific variance.}$$

With $i \neq j$, and $t = s$, the dependence is a contemporaneous spatial correlation, which may be the same for all time periods, or specific to each t. Formally:

$$E\ [\epsilon_{it} \epsilon_{jt}] = \sigma_{ij}\ (t), \qquad \text{contemporaneous correlation.}$$

With $i = j$, and $t \neq s$, the dependence is in the time domain, again, either constant over all spatial units, or varying with location (index i). Formally:

$$E\ [\epsilon_{it} \cdot \epsilon_{is}] = \sigma_{ts}\ (i), \qquad \text{time-wise correlation.}$$

When both $i \neq j$, and $t \neq s$, the pattern of dependence reaches across space and over time simultaneously:

$$E\ [\epsilon_{it} \cdot \epsilon_{js}] = \sigma_{ij}\ (ts), \qquad \text{space-time correlation.}$$

Clearly, as pointed out earlier, expression (4.7) is not operational, due to a lack of degrees of freedom to estimate β_{it} for every observation. The specific manner in which constraints are imposed on the general form allows for several interesting space-time modeling situations.

The exact opposite of (4.7) in terms of coefficient variability is:

$$y_{it} = x_{it}\beta + \epsilon_{it} \qquad (4.10)$$

in which the parameters are fixed across all observations in space and time. Situations of interest from a spatial modeling perspective occur when particular patterns of dependence or heterogeneity are present in the error term, similar to the categories just discussed.

Two broad classes of models are obtained when the β_{it} parameter is held fixed in one of the two dimensions. In both instances, a seemingly unrelated regression (SUR) situation results when the error terms are correlated across the other dimension.

With coefficient varying across space, but constant over time, the familiar Zellner (1962) example follows when the error terms are correlated contemporaneously:

$$y_{it} = x_{it}\ \beta_i + \epsilon_{it} \qquad (4.11)$$

with

$$E\ [\epsilon_{it} \cdot \epsilon_{jt}] = \sigma_{ij}.$$

In addition, residual autocorrelation (in the time dimension for each location i) is introduced if also

$$E\ [\epsilon_{it}.\epsilon_{is}] = \sigma_{ts}.$$

Alternatively, in the situation where the coefficient is specific to each time period, and constant across space, a different Zellner−type situation is obtained when the errors are correlated over time. This is often referred to as spatial SUR:

$$y_{it} = x_{it}\ \beta_t + \epsilon_{it}, \tag{4.12}$$

with

$$E\ [\epsilon_{it}.\epsilon_{is}] = \sigma_{ts}.$$

Introduction of spatial residual correlation in the cross−section for each time period follows from:

$$E\ [\epsilon_{it}.\epsilon_{jt}] = \sigma_{ij}.$$

In the specification above, a spatially lagged dependent variable, or a time−wise lagged dependent variable can be included among the X, to yield a still more complex class of models. Similarly, heteroskedasticity can be introduced by making the error variance different over time periods in model (4.11), and different over spatial units in model (4.12).

A particularly interesting specification issue consists of the choice between a general form or a specific parameterization for the spatial or time−wise dependence in the error terms. For example, in a spatial econometric context, using model (4.11), the contemporaneous correlation can be expressed in the form of a general covariance σ_{ij}, or can be parameterized as:

$$\epsilon = \rho W \epsilon + \mu$$

where, for each time period, the ϵ are a vector of error terms across space.

In contrast to the pure cross−sectional situation, where this reparameterization is a necessity, in space−time the choice between the unstructured and structured form for dependence can be based on other considerations.

Special forms for the error covariance also result when heterogeneity is expressed in terms of random coefficient variation or error component models. In the former, the variation of the coefficient β_{it} is formally expressed as:

$$\beta_{it} = \beta + \mu_{it}$$

where β is a mean value for the coefficients.

The error term allows for variation over space (i), time (t), or space−time. Consequently, in function of the assumptions about the μ_{it}, various patterns of error dependence and error heterogeneity can be encompassed within the general structure outlined above.

A similar situation exists for the error (or variance) component approach, where the individual error ϵ_{it} is taken to consist of a space, time and space–time component:

$$\epsilon_{it} = \mu_i + \lambda_t + \phi_{it}.$$

Again, specific assumptions about the dependence or independence of the three components result in particular forms for the error variance. These too are encompassed in the general specification outlined in this section.

NOTES ON CHAPTER 4

[1] For examples, see Whittle (1954), Besag (1974), Besag and Moran (1975), Bartlett (1975), and the overview in Haining (1979).

[2] See also Haining (1979), for details.

[3] For an extensive comparison of the simultaneous and conditional modeling strategies, see Haining (1977, 1978a, 1978b, 1979, and 1984). The last paper also provides an interesting application of the two types of models to a spatial pricing issue.

[4] See Bennett (1979) for an extensive overview.

[5] For a more extensive discussion of the properties of the spatial moving average model, see Haining (1978c).

[6] A major emphasis in the modeling of STARIMA specifications is the identification of the proper model structure and lag length from the data. As pointed out earlier, the data—driven approach will not be considered in detail in this book.

[7] Note that the exogenous variables may include spatial lags, such as WX. This is the case in the spatial Durbin model and the spatial common factors approach which will be discussed in Sections 8.2.2. and 13.3. The lagged exogenous variables can be treated in the same way as the other exogenous variables with respect to estimation and testing, although they may form a source of multicolinearity.

[8] Alternatively, the combination of ρ and the weight matrix could be expressed in a nonlinear form, as $W(d,\theta)$, where d represents a vector of distance (or accessibility) related variables, and θ is a vector of parameters.

CHAPTER 5

SPATIAL STOCHASTIC PROCESSES:

TERMINOLOGY AND GENERAL PROPERTIES

In this chapter I discuss in general terms the formal mathematical and statistical background for the estimation and testing of spatial econometric models. Even though this material is highly abstract, I will keep the treatment rather informal. An in—depth discussion of the various definitions, theorems and proofs would clearly be beyond the scope of this book. Since the relevant theoretical literature is extensive, a main motivation for this chapter is to bring together some important results that have been presented in a variety of sources.

The summary of some recent findings from the theoretical econometric literature emphasizes how these can be applied to general situations relevant in the treatment of spatial effects. Even though not developed with spatial applications in mind, these abstract results form the basis for the formal properties of various estimation procedures and specification tests for spatial models. In this respect, the theory of dependent and heterogeneous stochastic processes turns out to be extremely important.

The chapter consists of three sections. In the first, the notion of a spatial stochastic process is introduced and some important characteristics are briefly discussed. The second section summarizes several fundamental concepts from the literature on asymptotic approaches to estimation and inference, to the extent that these pertain to spatial and space—time processes. The chapter is concluded with a critical examination of the appropriateness of concepts such as convergence and asymptoticity for spatial modeling situations encountered in practice.

5.1. Properties of Spatial Stochastic Processes

5.1.1. Definitions

A stochastic process is a formal concept used to designate collections or sequences of random variables that are organized in some regular fashion. [1] More precisely, if a random variable, say x, can be associated with an index variable, say i, which pertains to a location in general p—dimensional space, R^p, then:

$$\{ x_i, i \in I \}$$

is a stochastic process. The range I, which is the collection of all possible values for the index parameter, can be any general dimensional space, discrete or continuous. A spatial stochastic process is the special case where the index pertains to a location in space. [2]

This family of random variables is characterized by a well—defined joint probability density or distribution. When this density is the normal, many simplifying results hold, and the process is called gaussian.

A stochastic process is completely specified when the joint distribution function is known for any finite subset of x_i which belong to the process. This necessitates a completely specified set of dependence relations between the x_i, $\forall$ i, a well—defined state space (i.e., the set of all possible values for the random variable) and index range. [3]

The notion of a spatial stochastic process is very different from the more traditional random sampling approach taken in empirical work. Indeed, the dependence of the values observed at different points in space is assumed to be generated by an unknown underlying process. The specific form of this process is determined a priori, presumably based on substantive theoretical grounds. In contrast, in the random sampling case, any dependence between the observational units is an unwanted feature. The independence is needed to infer information about a population from a subset of its members.

5.1.2. Stationarity and Isotropy

The stochastic process approach to spatial data means that essentially only one observation is available, i.e., the allocation of values to the random variables in space or space—time. Since this is not an operational situation, some restrictions need to be imposed on the degree of dependence and heterogeneity that can be allowed. Essentially, in order to infer certain characteristics of the underlying process, a degree of stability needs to be assumed. This is usually achieved by imposing the requirements of stationarity, ergodicity and isotropy.

A stationary stochastic process has reached a state of statistical equilibrium so that the generating mechanism can be assumed to work uniformly over space and time. In other words, the underlying joint distribution is taken to be the same for any subset of observations (homogeneity), which implies a number of restrictions on the moments, such as a constant mean and a constant finite variance.

In more formal terms, strict stationarity can be defined as the state where any finite subset,

$$\{ x_i, x_j, ..., x_n \}$$

from the stochastic process

$$\{ x_i, i \in I \}$$

has the same joint distribution as the subset

$$\{ x_{i+s}, x_{j+s}, ..., x_{n+s} \}$$

for any s, where s represents a uniform shift in time, space or space—time.

In a time series context, the notion of a shift is straightforward. Stationarity implies that the distribution of any x_i is the same at any point in time, that mean and variance are constant and finite, and that the covariance between x_i and x_j does not depend on the particular time period, but only on the shift in time (time lag). A weaker notion of stationarity, and the one used in most applications is covariance stationarity, where the time independence is required only for the moments up to order two.

The stationarity requirement in time series analysis leads to the use of autocovariance and autocorrelation functions indexed by the time shift, to aid in the identification and estimation of the models, as in the well—known approach pioneered by Box and Jenkins (1976).

In addition, for the underlying stochastic process to be indentifiable, and to ensure the existence of desirable asymptotic properties, some restrictions have to be imposed on the extent of the dependence. For example, one often used constraint consists of the requirement of ergodicity, which ensures that, on average, two events will be independent in the limit. [4]

In spatial analysis, the notion of a uniform shift is more complex, as already pointed out in the discussion of spatial weight matrices. In a strict sense, a spatial process is stationary when any joint distribution of the random variable over a subset of points depends only upon the relative position of the different locations, as determined by their relative orientation (angle) and respective distances. Since the orientation between points in two (or more) dimensions still leaves a great number of different situations (potentially over a 360 degree rotation), the stricter notion of isotropy is usually imposed as well. For an isotropic process, the joint distribution depends on the inter—location distance only, and orientation is irrelevant.

Again, weak or covariance spatial stationarity and isotropy can be defined by confining these requirements to first and second order moments. Consequently, for a weak spatial stationary process, the covariance between the random variable at two different locations depends only on the distance between the locations and the relative orientation with respect to a coordinate system. For an isotropic process, these moments depend only on the distance between two locations. In analogy to the time series situation, the analysis of spatial autocovariance and autocorrelation functions, indexed by the spatial shift, could form the basis for model identification and estimation. [5]

Hooper and Hewings (1981) have shown that this analogy is only appropriate in a limited class of spatial processes. Without the restrictive assumptions, no concept of spatial lag is available that provides a rigorous interpretation for the notion of stationarity and the associated use of spatial autocorrelation functions. Consequently, a meaningful treatment of spatial data (and space—time data) analogous to the Box—Jenkins approach in time series analysis is limited to these special processes. Unfortunately, the structure of spatial processes that obtain well—defined autocorrelation properties is highly restrictive. In addition, the nature of spatial stationarity and isotropy itself seems rather unrealistic for the type of data and spatial configurations encountered in applied regional science. This problem is not limited to spatial processes, but is relevant in the time domain as well. For example, objections to the unquestioning use of the assumption of stationarity in the analysis of economic time series are discussed in White and Domowitz (1984).

5.1.3. Locally Covariant Random Fields

A different way of structuring a spatial stochastic process such that meaningful estimation and hypothesis testing can be carried out is to restrict the degree of dependence explicitly. In the context of developing a central limit theorem for spatial samples (see also Section 5.2.3), Smith (1980) suggested the notion of a regular locally covariant random field.

In essence, this concept assumes that the spatial interaction (for homogeneous units) occurs strongly within a given radius (a neighborhood), while disappearing when the spatial units are far enough apart. It implies a particular form of contiguity, similar to the definitions considered in Section 3.1. In addition, the degree of heterogeneity is limited by imposing regularity conditions on the moments.

In formal terms, Smith defines a regular random field Ω as a spatial stochastic process which conforms to the two requirements of uniform nondegeneracy and Lyapunov boundedness. These are expressed as:

$$\sigma^2(\omega_i) \geq \delta \qquad \text{(uniform nondegeneracy)}$$

$$E(|\omega_i - \mu_i|^{2+d}) \leq \beta \qquad \text{(Lyapunov boundedness)}$$

The first requirement imposes a minimum degree of deviation from the mean (all variances positive and larger than some small value), while the second limits the extent of variation (no absolute moment higher than two should be larger than some large positive value β).

With a proper metric for expressing the distance between two spatial units of observation, the requirement of local covariation is defined as the combination of two postulates, adjacent covariation and remote independence. The first posits that for two spatial units that are close enough together (i.e., the distance between them is less than a specified positive value ϵ) positive covariation exists, i.e.,

$$E[\omega_i.\omega_j] > 0 \text{ for } d_{ij} < \epsilon.$$

On the other hand, when the spatial units (regions) are separated by a distance larger than ϵ, no dependence is allowed. This is the requirement of remote independence. [6]

The resulting structure forms a general framework within which formal statistical properties can be derived. The use of a distance metric as the basis for structuring space is similar to its importance in spatial interaction theory. Although this would imply a certain degree of isotropy, the flexibility of defining distance in general space essentially circumvents this limitation. Furthermore, the notion of a locally covariant random field is not limited to lattices, but can be applied to any arrangement of spatial units on which a regular distance metric can be defined. On the other hand, the requirement of positive covariance, and the degree of homogeneity that needs to be assumed may still be too restrictive for applied spatial models. [7]

5.1.4. Mixing Sequences

The concept of mixing sequences forms the most powerful framework to deal with dependence and heterogeneity in spatial processes. Although it is typically used to deal with dependence over time and heterogeneity across cross−sectional units, the general principles involved can be extended to spatial dependence in a fairly straightforward way.

Intuitively, the starting point for the notion of mixing is the distinction between two information sets that are subsets of the sample space (formally, σ−algebras, σ−fields, or Borel−fields). These information sets are separated by a distance m, in a suitably defined metric. In a time series context, this is expressed by two σ−algebras, say $\mathcal{G}$ and $\mathcal{H}$, where $\mathcal{G}$ represents all information in the past, from $-\infty$ to a given time n, and $\mathcal{H}$ represents all information in a *distant* future, from time n+m (with m as the distance) to $+\infty$. In space, the two information sets can be thought of in very general terms as pertaining to different regions or aggregates of spatial units, that are a given distance apart. Mixing coefficients provide a way to measure the dependence between the two information sets or σ−algebras.

In this formal setting, two measures of dependence between the information sets can be introduced, one expressing a relative measure (ϕ−mixing), the other an absolute measure (α−mixing). Formally, for two events G and H belonging to the respective information sets $\mathcal{G}$ and $\mathcal{H}$:

$$\phi\,(G,H) \equiv \sup\,\{G \in \mathcal{G}, H \in \mathcal{H},\ P(G){>}0\}\ |P(H|G) - P(H)|$$

$$\alpha\,(G,H) \equiv \sup\,\{G \in \mathcal{G}, H \in \mathcal{H}\}\ |P(G \cap H) - P(G){\cdot}P(H)|.$$

Clearly, if the two events G and H are independent, their joint probability equals the product of the marginal probabilities, and the conditional probability equals the unconditional marginal probability, so that both ϕ and α would be zero. A formal measure of the lack of independence between the two information sets is introduced when the ϕ and α coefficients are related to a measure of distance between the information sets (m). The concepts of ϕ−mixing and α−mixing coefficients formally express this as:

$$\phi\,(m) \equiv \sup_n\ \phi[\mathcal{B}(-\infty,n), \mathcal{B}(n{+}m, +\infty)]$$

$$\alpha\,(m) \equiv \sup_n\ \alpha[\mathcal{B}(-\infty,n), \mathcal{B}(n{+}m, +\infty)]$$

where $\mathcal{B}(-\infty, n)$ and $\mathcal{B}(n{+}m, +\infty)$ represent two information sets, separated by a distance m. A stochastic process is called α−mixing (or, strongly mixing) or ϕ−mixing (or, uniformly mixing) if the respective mixing coefficients approach zero as the distance m between the information sets increases to infinity:

α—mixing: $\alpha(m) \rightarrow 0$, with $m \rightarrow \infty$

ϕ—mixing: $\phi(m) \rightarrow 0$, with $m \rightarrow \infty$.

Although it is not always obvious what degree of mixing is implied by a particular stochastic process, the operational use of this concept is facilitated by the introduction of a formal link with moment restrictions. This results in an explicit trade—off between lack of independence (mixing) and allowable heterogeneity (moment restrictions) which forms the basis for a number of practical convergence properties and central limit theorems.

Intuitively, as the memory of a process decreases, i.e., with dependence rapidly decaying with increased distance, more heterogeneity can be allowed (restrictions on lower moments only). On the other hand, as the dependence increases in scope, tighter restrictions on higher order moments limit the extent of heterogeneity. Although, in some sense, the concept of mixing sequences is similar to the notions of stationarity and ergodicity, the explicit trade—off between dependence and heterogeneity allows for a much wider range of processes to be modeled and forms a framework with more realistic assumptions and restrictions. [8]

The extension of this general concept to spatial processes is straightforward, provided that the spatial units can be organized by means of a well—defined distance metric. Intuitively, distance over a time axis has an immediate analogy in distance over space. As long as the dependence over space decreases in some regular fashion with increasing distance, the analogy with mixing conditions can be drawn. This implies that the spatial weight matrix should reflect a distance decay pattern, where, as before, the weight matrix is considered to summarize a simultaneous pattern of spatial interaction. Weight matrices that are based on a formal concept of accessibility, such as in the axiomatic approach of Weibull (1976), would satisfy this requirement.

Since the spatial dependence is allowed to decay in a smooth fashion with increased distance, the mixing coefficient approach can be considered as a slightly more general framework than the abrupt notion of remote independence of Smith (1980). It also allows for a more satisfactory formal treatment of heterogeneity in space.

5.2. Asymptotic Approaches to Spatial Processes

The lack of independence present in spatial data sets precludes the application of much of the traditional sampling theory, which is based on the principle of independent random samples. Instead, the derivation of the formal properties of estimators and test statistics needs to be based on asymptotic considerations, i.e., on approximations which are valid when the number of observations in space, time, or space—time increases to infinity in a regular fashion. Intuitively, the loss of information due to the dependence in the sample is compensated by increasing the size of the data set. In a loose sense, the information contained in the larger dependent data set would be comparable to the information in a smaller collection of independent observations, provided that the scope of dependence does not increase in direct proportion to the number of observations.

The asymptotic approach to the statistical properties of estimators and tests is based on the viewpoint that the behavior of particular statistics (as functions of the data) which is not well understood in small samples, can be approximated by the behavior of other random variables (or constants) which is well understood. The latter typically leads to the use of the normal distribution, or some other distribution closely related to the normal, such as the χ^2. This approximation occurs as the size of the data set increases towards infinity. However, since the approximation strictly holds for very large data sets only, it does not necessarily provide a good guide for what happens in the actual finite sample.

Central limit theorems provide the formal structure and conditions under which random variables with unknown underlying distributions (e.g., functions of observations, including averages and cross—products computed from data) can be approximated in the limit by a normal variate with a known mean and variance. This forms the basis for *asymptotic hypothesis tests* in which inference about a population is obtained by using the normal variate as if it were the actual finite sample statistic. Clearly, this is not always appropriate.

Since the asymptotic approach is central in much of spatial econometrics, I will next briefly describe some fundamental concepts, important for the understanding of the properties of estimators and test statistics developed for the analysis of spatial processes.

5.2.1. Asymptotic Convergence

The way in which a sequence of random variables approaches a certain limit is governed by the concept of asymptotic convergence. Two notions are particularly important, convergence in probability and convergence in distribution. Both are extensions of the concept of limits to the domain of uncertainty. The first applies when a sequence of random variables approaches a constant real value or another random variable, the second when the distribution associated with the random variables in the sequence approaches another distribution function.

Loosely defined, convergence in probability is the situation where, in the limit, the probability is very small that a random variable in a sequence of random variables differs by a very small amount from a given constant, real value or from another random variable. Formally, a sequence of random variables x_i converges in probability to x, if for all $\epsilon > 0$:

$$\lim_{i \to \infty} P\left[|x_i - x| > \epsilon \right] = 0$$

The limit x is called the probability limit or *plim* of x_i. This notion is particularly relevant in the context of estimation, where an important property is consistency, i.e., when the *plim* of the estimator equals the population parameter.

A more stringent notion of convergence used in most central limit theorems is convergence almost surely (a.s.), where the probability of the limit of the difference between x_i and x is zero:

$$P[\ \lim_{i \to \infty}\ |x_i - x| > \epsilon\ ] = 0$$

The difference between the two concepts is rather abstract. Essentially, convergence almost surely necessitates a zero probability for a joint occurence of a small positive difference between x_i and x, for all large i. For convergence in probability, this requirement only pertains to each individual occurence. It is therefore a weaker notion. [9]

Convergence in distribution occurs when the distribution associated with the random variables in the sequence becomes arbitrarily close to a particular distribution function, for all values of the associated random variable. In formal terms, for all x in the domain of a distribution function $F_i(x)$, and for every $\epsilon > 0$, if there exists a (large) N such that:

$$|F_i(x) - F(x)| < \epsilon \qquad \qquad \forall\ i > N$$

at all continuity points of F, then x_i is said to converge in distribution to x, and $F(x)$ is called the limiting distribution of x_i. [10] This limiting distribution can be used to construct asymptotic tests of various hypotheses of interest.

5.2.2. Laws of Large Numbers

Laws of large numbers provide the formal conditions under which the average of a sequence of random variables converges to its expected value, for different types of underlying stochastic processes. These conditions form the basis for the convergence properties of various estimators and tests, which can be constructed from averages and weighted averages of the observations. The most familiar laws of large numbers deal with independent identically distributed variates, which is mostly irrelevant for spatial econometrics. More interesting are some recent results for dependent heterogeneous processes, based on the concept of mixing sequences discussed in the previous section.

The formal result is due to McLeish (1975), but a more applicable corrolary is given in White (1984, p. 47). It shows that almost sure convergence of the average of a series to the expected value of the average is obtained when the moments are bounded from above in function of the size of the mixing sequences. More precisely, almost sure convergence is obtained if

$$E\ |x_i|^{r+d} < D < \infty$$

for some d > 0 and $\forall$ i, and where the r are directly related to the size of the mixing coefficients. [11] In White (1984) and White and Domowitz (1984) it is shown how these abstract conditions can be applied to least squares estimators in models which contain lagged dependent variables as well as serially correlated and heteroskedastic error terms. [12]

5.2.3. Central Limit Theorems for Spatial Processes

Central limit theorems form the foundation for the derivation of limiting distributions for estimators and test statistics. Again, the more familiar results pertain to independent and identically distributed random variables, which is of little interest in spatial econometrics. However, some recent findings extend these results to situations where the underlying process can be dependent and heterogeneous.

In general terms, the central limit theorems pertain to sums of observations, which in turn are the building blocks for estimators that consist of sums of cross products of variables and error terms. In a regression context, these cross products are mostly of the form $X'\epsilon$, where the X are vectors of observations on explanatory variables and the ϵ are random error terms. At first sight, the properties of these cross products would depend on the assumptions for the underlying distribution of ϵ. The main contribution of the various central limit theorems is to establish asymptotic distribution results without the need to specify the exact underlying distribution, but only by imposing certain general restrictions on the degree of dependence and heterogeneity.

More formally, for a sequence of random variables x_i, with

$z_N = (1/N).\Sigma_i\, x_i$ as the average (for N observations)

$\mu_N = E\,[z_N]$ as the expected value of the average

$\sigma_N^2/N = \text{var}\,[z_N]$ as the variance of the average

a central limit theorem expresses the property that, given some restrictions on the dependence, heterogeneity and moments of x_i, but without specific assumptions about the underlying distribution,

$$N^{1/2}\,(z_N - \mu_N)/\sigma_N$$

is asymptotically distributed as a standard normal variate.

Of course, for these theorems to be operational, specific expressions for the expected value and variance have to be derived for each situation under consideration.

As already mentioned in Section 5.1.3, Smith (1980) extended the basic central limit result to spatial processes of a particular form, i.e., regular locally covariant random fields. Other results by Serfling (1980) and White and Domowitz (1984) have yielded central limit theorems for mixing processes, which include dependence and heterogeneity of a very general nature.

The existence of these central limit results for processes that are general enough to include most models of interest in spatial analysis provides a rigorous foundation for the asymptotic properties of estimators and tests. Of course, this generality comes at the cost of asymptoticity. Since finite sample results are not generally available for situations that incorporate substantial spatial dependence and/or heterogeneity, it is important to assess what these notions of limits and infinity mean for data situations typically encountered in applied regional science. I briefly turn to this issue next.

5.3.   A Closer Look at the Relevance of Statistical Approaches to Data in Space and Space—Time

The use of data organized by observational units in space and in space—time has given rise to some fundamental questions about the appropriateness of the application of statistical methods. Two issues in particular should be addressed: [1] the extent to which spatial data constitute a sample or form the complete population of interest; [2] the notion of convergence to infinitely large data sets in space, and its implications for statistical inference.

The sample vs. population issue has received some attention in the geographical literature, e.g., in Summerfield (1983), Openshaw (1984), and Johnston (1984). On the other hand, the issue associated with the use of an asymptotic framework have been mainly ignored. Both issues are intimately related to the extent to which spatial dependence and heterogeneity affect the amount of information that can be extracted from a given data set. I now consider them in turn.

5.3.1. Spatial Sample or Spatial Population

The models typically considered in spatial econometrics, in the sense in which it has been defined for the purposes of this book, are framed in the context of regression analysis. The two main objectives are inference and forecasting. The focus for the former is on determining the extent to which a theoretical relationship can be substantiated by empirical evidence, i.e., on estimating parameters and testing hypotheses. In forecasting, the interest lies in using empirically established relationships to predict values for unobserved observational units, i.e., forecasting future values, or interpolation in space.

Clearly, the imperfect state of theoretical knowledge and the imprecision in measuring the variables of interest are solid arguments for the inclusion of a stochastic error term. Furthermore, the stochastic process approach exemplifies the interest in the analysis of a sample to distinguish characteristics of an underlying unknown population. Based on this, a spatial econometric methodology should be considered to be relevant for the analysis of spatial data.

A contrasting view posits that cross—sectional data sets do not constitute a sample, but the whole population, and that therefore only a descriptive approach has merit, e.g., as argued by Summerfield (1983). Although this argument may seem to have intuitive appeal, it actually confuses a number of issues.

This confusion arises in part because of the limitations imposed by spatial dependence and spatial heterogeneity. In particular, when the heterogeneity is such that each spatial unit has its own unique characteristics, a cross—sectional analysis will not provide sufficient information to extract these. This does not imply that an econometric methodology would be invalid, but rather that more data (e.g., space—time data) is needed for it to be meaningfully applied.

Similarly, the property that spatial dependence leads to the inappropriateness of the traditional random sampling approach does not invalidate estimation and hypothesis testing. Instead, the focus should be on the extent to which the notion of an infinitely large sample can be applied to the observational context at hand. The essence of the problem is whether the information in the data set is compatible with the complexity of the model under consideration. Clearly, if this is not the case, more data is needed, rather than a rejection of an econometric or statistical methodology.

5.3.2. Asymptotics in Space and Space—Time

When data are used for a cross section of irregular spatial units, such as counties, states or provinces, the meaning of a notion of asymptoticity is not always clear. This is in contrast to the situation for a regular infinite lattice structure. In essence, the spatial units of observation should be representative of a larger population, and the number of units should potentially be able to approach infinity in a regular fashion. Clearly, this is not always immediately obvious for the type of data used in applied empirical work in regional science.

For example, consider the case where a multiregional econometric model is implemented for a given set of regions, such as all the counties in a state. The regions included in the data presumably have characteristics which set them apart from a larger population. This empirical situation can easily lead to considerable heterogeneity in the sample. In the extreme case, where each region has its own characteristics (own variables, own parameter values or own functional forms), there is clearly a need to structure this variety before meaningful statistical inference can be carried out. The same applies to potential spatial dependence between the regions in the sample.

There are two ways in which this can be approached. On the one hand, information over many time periods may be introduced, which, under the assumption that the patterns of spatial heterogeneity and dependence remain constant over time, allows for their identification and estimation. This is the usual approach taken for multiregional econometric models. Consequently, the formal properties of the estimators and tests are based on the time—dimension in the data. Also, the precise nature of dependence and heterogeneity can be left unspecified.

Alternatively, the heterogeneity itself can be structured as a specific function of space (i.e., reparameterized in function of spatial variables), for which the data set under consideration can be taken as a representative sample. This would be the situation where variables such as county area or county population would capture the heterogeneity between counties. The statistical properties could then be based solely on the cross—section dimension, provided that the sample could potentially be extended to include infinitely many *similar* counties.

When spatial dependence is present, the situation is more complex. In essence, the pattern of dependence in the sample should be assumed to be representative of a pattern in a hypothetical infinitely large set of contiguous spatial units. This necessitates that the spatial nature of the dependence is general enough to be applicable in this infinitely large data set. When the dependence is defined in terms of a well—behaved distance metric, this would not be a problem. In contrast, the extension of an ad hoc notion of first order contiguity to an infinitely large irregular spatial configuration is not necessarily meaningful. The general scope of the spatial dependence also implies that observations for regions outside the sample, but close enough to influence the regions that are included, may affect estimation and testing. This leads to the boundary value problem in spatial analysis, which is discussed more extensively in Section 11.2.

To conclude this chapter on the formal properties of spatial processes, it is important to stress the need for caution before a spatial econometric analysis is carried out for irregular spatial data sets. It is highly advisable to first assess whether the complexity of the proposed models is compatible with the limitations of the data. For many interesting models, there will simply not be enough information in a cross—sectional data set to allow for a meaningful analysis, and the use of space—time data is mandated. On the other hand, if the data set is of the kind which fits within the formal requirements discussed in this chapter, the econometric methodology will be based on a rigorous probabilistic foundation.

NOTES ON CHAPTER 5

[1]   A random variable is defined as a set theoretic function that associates a real number with the outcome of an uncertain event. Formally, for a given probability space $(\Omega, \mathcal{F}, P)$, which is defined in terms of a sample space Ω (the collection of all possible outcomes ω for an uncertain event), a Borel algebra or σ-algebra $\mathcal{F}$ (a completely additive class of sets), and a probability measure P, a random variable ω is a function with domain Ω and counterdomain the real line, for which: $A_r = \{ \omega : x(\omega) \leq r \} \in \mathcal{F}, \forall\, r \in \Re$ (where $\Re$ is the set of real numbers). This axiomatic and measure theoretic approach to probability is common to all recent treatments in mathematical statistics. Specifically, the use of the notion of a Borel algebra or σ-algebra allows for the consideration of infinite sequences of subsets of the sample space. This forms the basis for the main results in the asymptotic theory of econometrics. More precisely, a Borel or σ-algebra $\mathcal{F}$ is a class of sets, for which: i. the sample space $\Omega \in \mathcal{F}$; ii. if a subset of the sample space belongs to $\mathcal{F}$, then its complement also belongs to $\mathcal{F}$; and iii. if A_1, A_2, ... is a countably infinite sequence of sets, with each $A_i \in \mathcal{F}$, then their (infinite) union is also an element of $\mathcal{F}$ (this extends the notion of a Boolean algebra or Boolean field of sets to infinitely large sample spaces). Random variables and functions of random variables are considered to be measurable with respect to a particular σ-algebra (i.e., a organized subset of events). For a more extensive and rigorous discussion of these concepts, see, e.g., Wilks (1962), Mood, Graybill and Boes (1974), Chung (1974), Pfeiffer (1978), and Billingsley (1985).

[2]   In Smith (1980), this is referred to as a p-dimensional random field, for which the range is taken as a p-dimensional euclidean space. For alternative overviews and similar definitions, see, e.g., Bartlett (1975), and Haining (1977, 1979).

[3]   For a more extensive discussion of the notion of a stochastic process, see, e.g., Karlin and Taylor (1975), Box and Jenkins (1976), Bartlett (1978), and Bennett (1979).

[4]   In a time series context, the ergodicity requirement consists of a limit on the *memory* of the sequence, and boils down to a form of asymptotic independence. More precisely, the independence is between two events. One is expressed as F, and the the other is a measure preserving transformation of an event G, denoted as a shift S over s steps, $S^s(G)$. The size of the shift ranges from the smallest possible one (e.g., one time period) to the full sample size N. Formally, the asymptotic independence can be expressed as:

$$\lim_{N \to \infty} N^{-1} \sum_s P[F \cap S^s(G)] = P(F) \cap P(G)$$

See, e.g., Doob (1953), Karlin and Taylor (1975), and White (1984) for a more rigorous discussion of this issue.

[5]   For some early discussion, see Granger (1969) and Berry (1971). More recent overviews of these issues are given in Bennett (1979), Cliff and Ord (1981), and Ripley (1983).

[6]   Intuitively, the notion of remote independence is similar to the concept of ergodicity in a time series context. Smith (1980, p. 303) also introduces the notion of a dispersed spatial sampling scheme, which ensures a spatial configuration in which no single region (as an aggregate of individual spatial units) contains arbitrarily many points.

[7]   Smith (1980) points to some possible extensions which would allow negative covariance as well. However, these are more appropriately included in the mixing processes discussed in Section 5.1.4.

[8]   An excellent discussion of the relevance of the various concepts of mixing sequences for econometric time series and panel data is given in Domowitz and White (1982), White (1984), and White and Domowitz (1984). Earlier, more formal sources are Rosenblatt (1956) and Ibragimov and Linnik (1971). Smith (1980, p. 313) refers to the concept of α-mixing in his formal proof for a central limit theorem for spatial processes, but does not implement this notion in the sense of a smooth variation with distance. In Anselin (1984b, 1986a) the concept of mixing sequences is used as a motivation for extending econometric procedures developed for time series analysis to the spatial domain.

[9]   Since concepts of convergence are related to a measure of distance between the sequence and its limit, various definitions are possible, depending on the notion of distance used. A well-known example is convergence in quadratic mean of vectors of random variables. A detailed discussion of formal concepts of convergence is beyond the current scope. Extensive and rigorous treatments of this issue can be found in, e.g., Theil (1971, Chapter 8), Serfling (1980, Chapter 1), Greenberg and Webster (1983, Chapter 1), and Amemiya (1985, Chapter 3).

[10]   The limiting distribution has a mean, which is called the asymptotic expectation of x_i. However, this is not the same as the *plim* of the mean of x_i. For details, see the references cited in the previous Note.

[11]   In White (1984), the size of a mixing coefficient is formally defined as a function of a real number r, with $1 \leq r \leq \infty$. For ϕ mixing, if $\phi(m)$ is at most of order m^{-k}, for $k > r/(2r-1)$, then $\phi(m)$ is said to be of size $r/(2r-1)$ (a sequence $\{x_N\}$ is at most of order N^k, or $O(N^k)$, if there exists a real number M, such that $N^{-k}.|x_N| \leq M$, $\forall$ N). Also, if $r>1$ and $\alpha(m)$ is at most of order m^{-k}, for $k>r/(r-1)$, then $\alpha(m)$ is said to be of size $r/(r-1)$. The trade-off between dependence and heterogeneity is expressed in the relation between moment conditions in terms of r and the magnitude of the mixing coefficients. Higher order conditions (large r) are necessary as the sequence shows more dependence.

[12]   A more extensive treatment of these formal properties is beyond the scope of this book. The interested reader is referred to the rigorous and comprehensive discussion in White (1984), and the references cited therein, as well as to the specific applications in Domowitz and White (1982) and White and Domowitz (1984).

PART II

ESTIMATION AND HYPOTHESIS TESTING

CHAPTER 6

THE MAXIMUM LIKELIHOOD APPROACH

TO SPATIAL PROCESS MODELS

In this chapter, I consider the application of the maximum likelihood principle to estimation and hypothesis testing for spatial process models. The models follow the taxonomy for cross-sectional situations presented in Chapter 4. Space-time formulations will be discussed in Chapter 10.

The maximum likelihood estimation of various spatial models has received considerable attention in the literature. For example, formal derivations for various models are discussed at length in several recent spatial statistics texts, such as the ones by Cliff and Ord (1981), Ripley (1981), and Upton and Fingleton (1985). In order to avoid duplication, I will not present much detail on these better-known derivations. Instead, the focus in this chapter is on some issues and model formulations that have received less attention elsewhere.

Specifically, I use the comprehensive model presented in Chapter 4 to illustrate the derivation of a general likelihood approach. Most of the models discussed in the literature can be found as special cases of this encompassing formulation. I also focus considerable attention on a general approach to hypothesis testing in the maximum likelihood context. This encompasses the asymptotic Wald test, Likelihood Ratio principle, and Lagrange Multiplier approach. The latter in particular has not received much attention in the literature. Since all derivations are presented in matrix notation, some useful results on matrix differentiation are summarized in appendices.

I first briefly consider the limitations of the ordinary least squares estimator for spatial process models.

6.1. Limitations of Ordinary Least Squares Estimation in Spatial Process Models

The spatial dependence in the various spatial autoregressive models shows many similarities to the more familiar time-wise dependence. Therefore, one would expect the properties of least squares estimation for models with lagged dependent variables and/or serial residual correlation to translate directly to the spatial case. However, this is not so. The lack of a direct analogy is primarily due to the two-dimensional and multidirectional nature of dependence in space. Next, I illustrate this property for the situation of a spatially lagged dependent variable and the case of spatial autocorrelation in the regression residual.

6.1.1. OLS in the Presence of a Spatially Lagged Dependent Variable

It is a well known result in econometrics that the OLS estimator remains consistent even when a lagged dependent variable is present, as long as the error term does not show serial correlation. Consequently, even though the small sample properties of the estimator are affected (it is no longer unbiased) it can still be used as the basis for asymptotic inference.

For spatial autoregressive models, this result does not hold, irrespective of the properties of the error term.

Consider the following pure first order spatial autoregressive model:

$$y = \rho.W.y + \epsilon \tag{6.1}$$

where W is the usual spatial weight matrix, ρ is a spatial autoregressive coefficient, the y are expressed in deviations from the mean, and the ϵ is an independent identically distributed error term. Even though this model is extremely simple, it captures all the effects of the presence of a spatially lagged dependent variable on the OLS estimate, and can therefore be used without loss of generality.

The OLS estimate for ρ, denoted by r is:

$$r = (y_L'y_L)^{-1}y_L'y \tag{6.2}$$

with $y_L = Wy$ as the spatially lagged dependent variable.

Substituting the expression for y in the population parameters from (6.1) in (6.2) gives:

$$r = \rho + (y_L'y_L)^{-1}y_L'\epsilon$$

Similar to the situation in time series, the expected value of the second term does not equal zero, and therefore the OLS estimate is biased. [1]

Asymptotically, the consistency of the OLS estimator depends on the following two conditions:

$plim \ N^{-1} \ (y_L'y_L) = Q$, a finite and nonsingular matrix

$plim \ N^{-1} \ (y_L'\epsilon) = 0.$

Whereas the first condition can be satisfied with the proper constraints on the value of ρ and the structure of the spatial weight matrix, the second condition does not hold in the spatial case. Indeed,

$$plim \ N^{-1} \ (y_L'\epsilon) = plim \ N^{-1} \ \epsilon'W(I-\rho W)^{-1}\epsilon$$

The presence of the spatial weight matrix in this expression results in a quadratic form in the error terms. Therefore, except in the trivial case where $\rho=0$, the *plim* of this expression will not equal zero.

Consequently, the OLS estimator will be biased as well as inconsistent for the parameters of the spatial model, irrespective of the properties of the error term.

6.1.2. OLS in the Presence of Spatial Residual Autocorrelation

The effects of spatial residual autocorrelation on the properties of the OLS estimator are more in line with the time series results. Parameter estimates will still be unbiased, but inefficient, due to the nondiagonal structure of the disturbance variance matrix.

The spatial model can be considered as a special case of the linear regression model with a general parameterized variance matrix for the error term. Therefore, the usual properties of OLS and GLS will apply. However, in the spatial case, the multidirectional nature of the spatial dependence will limit the type of EGLS (Estimated Generalized Least Squares) procedures that will lead to consistent estimates. Specifically, this applies to the various two−step EGLS procedures commonly used in the case of serial autocorrelation and heteroskedasticity.

Based on the results from the previous section, applied to a spatial autoregressive structure in regression residuals, it can be shown that no consistent estimate for the autoregressive parameter can be obtained from ordinary least squares estimation. Consequently, a simple spatial analogue of the well−known Cochrane Orcutt procedure will not be appropriate. [2]

This issue is pursued in more detail in Chapter 8.

6.2.   Maximum Likelihood Estimation

The inappropriateness of the least squares estimator for models that incorporate spatial dependence has focused attention on the maximum likelihood approach as an alternative. Going back to the early work of Whittle (1954) and Mead (1967), maximum likelihood approaches have been suggested and derived for spatial autoregressive and spatial moving average models by Cliff and Ord (1973), Ord (1975), Hepple (1976), Hordijk and Paelinck (1976), Haining (1978a, 1978c, 1978d), Brandsma and Ketellapper (1979a), Anselin (1980), Doreian (1982), Cook and Pocock (1983), and Blommestein (1985), among others. [3]

Below, I outline the derivation of ML estimators and the associated asymptotic variance matrix for the general model presented in Section 4.2. This model includes a spatially lagged dependent variable, spatial autoregression in the disturbance term as well as heteroskedasticity of a specified form. Before proceeding with the detailed derivations, I first briefly consider the formal properties of ML estimators for this general model.

6.2.1. Properties of ML Estimators in the Presence of Spatially Lagged Dependent Variables

Typically, the usual attractive asymptotic properties of the maximum likelihood estimator are assumed to apply for the models with spatially lagged dependent variables without much consideration. However, it was not until very recently that formal conditions for the consistency, efficiency and asymptotic normality of ML estimates for dependent observations have been derived, in the work of Bates and White (1985), and Heijmans and Magnus (1986a, 1986b, 1986c). [4] In general terms, these conditions boil down to the following requirements: the existence of the log likelihood for the parameter values under consideration (i.e., a non−degenerate log−likelihood); continuous differentiability of the log likelihood (to the second or third order, and for parameter values in a neighborhood of the true value); boundedness of various partial derivatives; the existence, positive definiteness and/or non−singularity of covariance matrices; and the finiteness of various quadratic forms. [5]

A further requirement, important in the specification of spatial dependence and spatial heterogeneity, is that the number of parameters should be fixed and independent of the number of observations. This is to avoid the so−called incidental parameter problem, which is a common issue in space−time models. For example, a situation where each spatial unit would have a separate parameter to indicate its own particular features (e.g., a dummy variable) would not be admissible in a context where asymptoticity is based on the cross−sectional dimension. Similarly, a model where the spatial dependence for each observation would carry its own distance−decay parameter would be inappropriate. In applied situations, it is usually fairly obvious when this requirement is not fulfilled, from the lack of degrees of freedom.

For the spatial models considered here, the various conditions are typically satisfied when the structure of spatial interaction, which is expressed jointly by the autoregressive coefficient and the weight matrix, is nonexplosive. Formally, this can be assessed by studying the properties of the Jacobian associated with each model, e.g., det $(I-\rho W)$ in the simple spatial autoregressive formulation. [6] I return to this issue in more detail in Section 6.2.3.

From a practical point of view, it is usually quite obvious when the conditions are not satisfied, from a closer analysis of *problems*, such as the non−convergence of nonlinear optimization routines, singular or negative−definite variance matrices, and numerical overflow or underflow (e.g., division by zero of by infinitely large values).

6.2.2. Properties of the ML Estimator for the Linear Regression Model with Spatial Effects in the Error Terms

The linear model with spatial effects in the disturbance term can be considered as a special case of the model with a general nonscalar error covariance matrix. This model formulation has recently received considerable attention in the econometric literature. The covariance matrix is usually unknown and expressed in terms of a finite number of parameters. In the spatial model, these would be the autoregressive parameter(s), distance decay parameter(s), or the coefficient(s) associated with the variable(s) which specify the form for the heteroskedasticity.

When the disturbance variance matrix is known, the familiar GLS (Generalized Least Squares) estimator is BLUE (Best Linear Unbiased Estimator) as well as maximum likelihood. However, in the more realistic situation where the parameters of the error variance are based on an estimate, the properties of the resulting EGLS (or feasible GLS) are not immediately clear. Recent results in the econometric literature, e.g., by Magnus (1978), Rothenberg (1984a), and Andrews (1986), show that an ML estimator (as well as other EGLS approaches), under general conditions, achieves the desirable properties of consistency, asymptotic efficiency and asymptotic normality. Moreover, in most situations of interest, the resulting estimates for the regular parameters of the model (the β in the usual notation) are also unbiased.

The regularity conditions are similar to the ones discussed in the previous section, and are designed primarily to guarantee a well−behaved likelihood function and a positive definite error covariance matrix. For the spatial model, these conditions are essentially satisfied by a non−explosive structure of interaction in the spatial weight matrix, and by imposing non−negativity constraints on the diagonal elements of the estimated error covariance matrix, i.e., on the coefficients in the function for heteroskedasticity.

6.2.3. The Likelihood Function and Jacobian for the General Model

As in Section 4.2, the specification for the general spatial process model is:

$$y = \rho W_1 y + X\beta + \epsilon \tag{6.3}$$

$$\epsilon = \lambda W_2 \epsilon + \mu \tag{6.4}$$

with $\mu \sim N(0, \Omega)$, $\tag{6.5}$

and the diagonal elements of the error covariance matrix Ω as:

$$\Omega_{ii} = h_i(z\alpha) \qquad h_i > 0. \tag{6.6}$$

In all, the model has $3+K+P$ unknown parameters, in vectorform:

$$\theta = [\rho, \beta', \lambda, \sigma^2, \alpha']'. \tag{6.7}$$

This model can also be expressed in a non−linear form, which facilitates the illustration of the relevant results. I use the following simplification in notation:

$$A = I - \rho W_1$$

$$B = I - \lambda W_2,$$

which gives, for (6.3) and (6.4):

$$Ay = X\beta + \epsilon \tag{6.8}$$

$$B\epsilon = \mu \tag{6.9}$$

Also, since the error covariance matrix

$$E \left[\mu\mu'\right] = \Omega,$$

is diagonal, there exists a vector of homoskedastic random disturbances v, as

$$v = \Omega^{-1/2}\mu, \tag{6.10}$$

or, alternatively,

$$\mu = \Omega^{1/2}v, \tag{6.11}$$

and the disturbance in (6.9) becomes

$$\epsilon = B^{-1}.\Omega^{1/2}v \tag{6.12}$$

Substituting (6.12) in (6.8) gives:

$$Ay = X\beta + B^{-1}.\Omega^{1/2}v,$$

or, alternatively,

$$\Omega^{-1/2}.B.(Ay - X\beta) = v \tag{6.13}$$

In this nonlinear expression (i.e., nonlinear in the parameters), v is a vector of standard normal and independent error terms. Consequently, (6.13) conforms to the usual expression for the implicit form of nonlinear models,

$$f(y,X,\theta) = v$$

where f is a general nonlinear functional form relating y, X and a vector of parameters θ, and v is the disturbance term.

Although the error term v has a well behaved joint distribution, it cannot be observed, and the likelihood function has to be based on y. Therefore, it is necessary to introduce the concept of a Jacobian, which allows the joint distribution for the y to be derived from that for the v, through the functional relationship expressed in (6.13).

The Jacobian for the transformation of the vector of random variables v into the vector of random variables y is:

$$J = det \left(\partial v/\partial y\right)$$

which, using (6.13), becomes:

$$|\Omega^{-1/2}.B.A| = |\Omega^{-1/2}|.|B|.|A| \tag{6.14}$$

Based on a joint standard normal distribution for the error term v, and using (6.14), the log−likelihood function for the joint vector of observations y is obtained as:

$$L = -(N/2).\ln(\pi)-(1/2).\ln|\Omega|+\ln|B|+\ln|A|-(1/2)v'v \tag{6.15}$$

with,

$$v'v = (Ay - X\beta)'B'\Omega^{-1}B.(Ay - X\beta) \tag{6.16}$$

as a sum of squares of appropriately transformed error terms.

From (6.15) it follows that a maximization of the likelihood function is equivalent to a minimization of a sum of squared (transformed) errors, corrected by the determinants from the Jacobian. This correction, and in particular its spatial terms in A and B, will keep the least squares estimate from being equivalent to ML. The extent of the difference between the two estimators is largely a function of the magnitude of these two determinants. Specifially, for standardized weight matrices W_1 or W_2, as either $\rho \to +1$ or $\lambda \to +1$, the adjustment becomes infinitely large. [7]

The essential part of the log−likelihood consists of a quadratic form in the error terms, which leads to a well−behaved optimization problem. However, the determinants $|\Omega|$, $|A|$ and $|B|$ in (6.15) can cause problems in this respect. Indeed, the asymptotic properties for the ML estimates will only hold if the regularity conditions for the log−likelihood function are satisfied. In the current context, both A and B can lead to explosive behavior for particular parameter values, and Ω may fail to be positive definite. [8]

It is therefore necessary to ensure that the following general condition holds for the Jacobian:

$$|\Omega^{-1/2}.A.B| > 0 \tag{6.17}$$

which is satisfied by the partial requirements: [9]

$$| I - \rho.W_1 | > 0 \tag{6.18}$$

$$| I - \lambda.W_2 | > 0 \tag{6.19}$$

$$h_i (z\alpha) > 0, \forall i. \tag{6.20}$$

The constraint (6.20) is a familiar one in random coefficient models (see also Section 9.4.1.). Constraints (6.18) and (6.19) result in restrictions on the values that the spatial autoregressive coefficients can take. For standardized weight matrices this usually means that the parameter should be less than one. [10]

6.2.4. The First Order Conditions for ML Estimates in the General Model

The first order conditions for the ML estimators in model (6.13) are obtained by taking the partial derivatives of the log−likelihood (6.15) with respect to the parameter vector. This involves a tedious but fairly straightforward application of matrix calculus, which is discussed in some detail in Appendix 6.A. to this chapter.

The resulting vector of first partial derivatives, the score vector, is set equal to zero and needs to be solved for the parameter values:

$$d = (\partial L/\partial \theta) = 0$$

with as elements of d: [11]

$$\partial L/\partial \beta = v'(\Omega^{-1/2}.B.X) \tag{6.21}$$

$$\partial L/\partial \rho = - \text{ tr } A^{-1}W_1 + v'\Omega^{-1/2}.B.W_1.y \tag{6.22}$$

$$\partial L/\partial \lambda = - \text{ tr } B^{-1}W_2 + v'\Omega^{-1/2}.W_2.(Ay-X\beta) \tag{6.23}$$

$$\partial L/\partial \alpha_p = - (1/2) \text{ tr } \Omega^{-1}.H_p + (1/2) \ v'\Omega^{-3/2}.H_p.B.(Ay - X\beta) \tag{6.24}$$

for p=1,...,P.

Clearly, this system of highly nonlinear equations does not have an analytic solution and needs to be solved by numerical methods. For the special spatial models included in the general specification (6.13), i.e., the first order spatial autoregressive and mixed regressive autoregressive model, and the linear regression model with spatial autoregressive errors, this is not too complex. Indeed, part of the first order conditions have a solution which can be used to construct a concentrated likelihood function, which is nonlinear in only one parameter. [12] Similarly, the situation where only heteroskedasticity is present results in a less complex nonlinear optimization problem. [13] The operational implementation of the various maximum likelihood estimators is further discussed in Section 12.1. All empirical illustrations for the estimators and tests are presented in Section 12.2.

6.2.5. The Asymptotic Variance Matrix for the General Model

Under the usual regularity conditions, the ML estimates that are found as solutions to the system (6.21)−(6.24) will be asymptotically efficient. This means that they achieve the Cramer−Rao lower variance bound, given by the inverse of the information matrix:

$$[I(\theta)]^{-1} = - \text{ E } [\partial^2 L/\partial \theta \partial \theta']^{-1}$$

The elements of the information matrix are found by taking the second partial derivatives with respect to the elements of the parameter vector θ, and by using the structure for the disturbance terms given in (6.4) and (6.9)−(6.12) to derive the relevant expected values. A detailed derivation is presented in Appendix 6.B.

For the various combinations of parameters, the following results are obtained:

$$I_{\beta\beta'} = X'B'\Omega^{-1}BX \tag{6.25}$$

$$I_{\beta\rho} = (BX)'\Omega^{-1}.B.W_1.A^{-1}.X\beta \qquad (6.26)$$

$$I_{\beta\lambda} = 0 \qquad (6.27)$$

$$I_{\beta\alpha'} = 0 \qquad (6.28)$$

$$I_{\rho\rho} = tr(W_1.A^{-1})^2 + tr\Omega.(B.W_1.A^{-1}.B^{-1})'\Omega^{-1}(B.W_1.A^{-1}.B^{-1})$$

$$+ (B.W_1.A^{-1}.X\beta)'\Omega^{-1}(B.W_1.A^{-1}.X\beta) \qquad (6.29)$$

$$I_{\rho\lambda} = tr(W_2.B^{-1})'\Omega^{-1}.B.W_1.A^{-1}.B^{-1}.\Omega + tr \; W_2.W_1.A^{-1}.B^{-1} \qquad (6.30)$$

$$I_{\rho\alpha_p} = tr\Omega^{-1}.H_p.B.W_1.A^{-1}.B^{-1} \qquad (6.31)$$

$$I_{\lambda\lambda} = tr(W_2.B^{-1})^2 + tr\Omega.(W_2.B^{-1})'\Omega^{-1}.W_2.B^{-1} \qquad (6.32)$$

$$I_{\lambda\alpha_p} = tr\Omega^{-1}.H_p.W_2.B^{-1} \qquad (6.33)$$

$$I_{\alpha_p\alpha_q} = (1/2) \; tr\Omega^{-2}.H_p.H_q \qquad (6.34)$$

The asymptotic variance matrix is obtained by substituting the ML estimates for the parameters in expressions $(6.25)-(6.34)$ and taking the inverse of the information matrix. Since the dimension of this matrix is $3+K+P$, no analytic results are available. [14]

The estimated asymptotic variance matrix can then be used as the basis for various hypothesis tests. This is discussed in the next section.

6.3. Hypothesis Tests Based on the Maximum Likelihood Principle

Given the widespread use of the maximum likelihood approach in the estimation of spatial process models, most hypothesis tests for the parameters of these models are based on asymptotic considerations as well. Of the three familiar asymptotic testing principles, i.e., the Wald test (W), Likelihood Ratio test (LR), and the Lagrange Multiplier test (LM), the first two have received most of the attention in spatial econometrics. Even though LM tests are sometimes referred to, they are seldom carried out.

Most of the inference in spatial models is based on the Wald (asymptotic t-test) or LR test, as illustrated by the general emphasis in Cliff and Ord (1973, 1981), Brandsma and Ketellapper (1979b), Anselin (1980), and Upton and Fingleton (1985). Exceptions are the tests suggested in Burridge (1980, 1981), and more recently, the comprehensive approach in Anselin (1988a), which are based on the LM principle.

In contrast to the extensive treatment of ML estimation, the formal properties of the various testing approaches have not received much attention in the spatial econometric literature. Moreover, the discussion is often rather imprecise and without much regard for the implications of the use of these tests in finite samples. In this section, I will therefore focus on these issues somewhat more in depth. The emphasis will be on testing hypotheses on general functions of the model parameters. A specific treatment of spatial effects in the error term of the linear regression model is relegated to Chapter 8.

6.3.1. General Principles

The Wald, LR and LM tests are based on the optimal properties of the maximum likelihood estimator. More specifically, these properties follow from the asymptotic normality of estimates and of functions of estimates. [15]

Formally, with the ML estimate for the vector of parameters θ as h:

$$N^{1/2} (h-\theta) \rightarrow N[0, \lim_{N\to\infty} (I(\theta)/N)^{-1}]$$

i.e., the difference between the estimate and the population parameter converges in distribution to a normal distribution with a zero vector as mean, and a variance which corresponds to the inverse information matrix.

Most hypothesis tests can be formulated as tests on functions of the model parameters:

$$H_0 : g(\theta) = 0$$

$$H_1 : g(\theta) \neq 0$$

where g is a q-dimensional linear or nonlinear matrix function in the elements of the parameter vector θ. As a special case, this includes tests on the significance of individual coefficients. In this instance, the function g simplifies to a vector with a one corresponding to the coefficient of interest, and zeroes elsewhere.

In the general spatial model, the focus is on the significance of either of the autoregressive processes (e.g., with H_0: $\rho=0$, or H_0: $\lambda=0$), on the regressive parameters β, and on tests for the presence of heteroskedasticity (e.g., with H_0: $\alpha=0$).

In general terms, the three asymptotic testing approaches are based on different measures of the distance between an unrestricted estimate and an estimate that satisfies the constraints implied by the null hypothesis, i.e., the restricted estimate. For example, if the vector of parameters θ is partitioned as $\theta' = [\theta_1' \mid \theta_2']$, the null hypothesis could be of the form:

$$H_0 : \theta_1=0.$$

A restricted estimate (θ_R) would consist of estimates for θ_2, with all parameters in θ_1 set to zero. The unrestricted estimate is the usual full vector θ. The tests would then be based on measures of the difference between the estimates for the full θ and the restricted θ_R. Intuitively, if the distance between the two results is too large, the restrictions cannot be taken to hold, and thus the null hypothesis is rejected.

The manner in which the tests measure the distance between restricted and unrestricted estimation also has implications for the kinds of estimates needed to carry out the respective tests. For the Wald test the full model needs to be estimated, i.e., unrestricted parameters are used. For the LM test restricted coefficients form the starting point, i.e., the model is estimated in a simpler specification with the null hypothesis imposed. For the LR test, both restricted and unrestricted estimation has to be carried out.

A more extensive and rigorous discussion of the salient properties and distinguishing characteristics of the three testing approaches can be found in, e.g., Breusch and Pagan (1980), Buse (1982), Engle (1982, 1984), and Davidson and MacKinnon (1983, 1984).

The most familiar of the three procedures is undoubtedly the Wald test, expressed in its most general form as:

$$W = g'[G'VG]^{-1}g$$

with

- g as a q by 1 vector of the values which result when the constraints are evaluated for the ML parameter estimates;
- G as a $(3+K+P)$ by q matrix of partial derivatives $\partial g'(\theta)/\partial \theta$, evaluated for the parameter estimates;
- V as the estimated asymptotic variance matrix, of dimension $3+K+P$.

The Wald statistic is asymptotically distributed as χ^2 with q degrees of freedom, where q corresponds to the number of constraints. Since the square root of the Wald test corresponds to a standard normal variate, it is asymptotically equivalent to a Student t-variate. In the context of testing the significance of the model parameters, the Wald test is therefore commonly (although somewhat misleadingly) referred to as an asymptotic t-test.

The LR test is based on the difference between the log-likelihood for the unrestricted (with θ) and the restricted model (with θ_R):

$$LR = 2[L(\theta) - L(\theta_R)]$$

where the L are the corresponding log-likelihoods. The LR test is also asymptotically distributed as χ^2 with q degrees of freedom.

The LM test, also known as score test, is based on an optimization approach, or more precisely, on the first order conditions for the optimization of a lagrangian function in the log-likelihood:

$$f = L(\theta) + \eta'g(\theta)$$

where f is the lagrangian, L is the log-likelihood, and η is a vector of lagrange multipliers corresponding to the q restrictions $g(\theta)$. The test itself turns out to be expressed only in the restricted coefficients, i.e., based on estimation in the simpler model:

$$LM = d_R'I(\theta_R)^{-1}d_R$$

with,

- d_R as the score vector for the encompassing model, $\partial L/\partial\theta$, evaluated at the null, i.e., with the constraints imposed;

- $I(\theta_R)$ is a consistent estimator for the information matrix, also evaluated at the null.

The LM test statistic is also distributed asymptotically as χ^2 with q degrees of freedom.

In most situations of interest in a regression context, the null hypothesis can be formulated as an omitted variable problem. In other words, with the parameter vector partitioned as $\theta' = [\theta_1' \mid \theta_2']$, H_0 is expressed as $\theta_1=0$. In this case, the operation on the information matrix simplifies to a partitioned inversion conforming to the partitioning in θ, and evaluated at the null. Moreover, in most situations, the resulting test statistic reduces to a simple expression in regression residuals, usually found in a straightforward way as functions of the R^2 in an auxiliary regression. However, it turns out that many of these simplifying results do not hold for spatial models. I return to this issue in more detail in Section 6.3.4.

The three tests can be shown to be asymptotically equivalent, although they will typically differ in finite samples. This is discussed more extensively in Section 6.3.5. First, I illustrate how the general principle can be applied to obtain tests on the parameters of spatial process models.

6.3.2. Wald Tests in Spatial Process Models

The tests that are likely to be of most relevance to applied spatial econometric work are those that pertain to the significance of individual model coefficients, or to the joint significance of the complete parameter vector. In both cases the parameter constraints are linear, which simplifies the expression for the Wald test.

As an example, consider a significance test on the spatial autoregressive parameter ρ in the spatial model. The corresponding constraints would be expressed as:

$$H_0 : [1\ 0'].[\rho,\ \beta',\ \lambda,\ \sigma^2,\ \alpha']' = \rho = 0$$

where $0'$ is a $2+K+P$ row vector of zeroes. Consequently, the partial derivative of the constraint with respect to ρ becomes the vector $[1\ 0']$, and the Wald test reduces to:

$$W = r.\{[1\ 0']V[1\ 0']'\}^{-1}.r$$
$$= r^2/v_{11} \sim \chi^2 \quad (1)$$

with r as the ML estimate for ρ, and v_{11} as the diagonal element corresponding to ρ in the parameter variance matrix. Clearly,

$$r/se_{11} \sim N(0,1)$$

with se_{11} as the estimated standard deviation.

Similarly, a joint significance test on all model parameters in the general formulation would consist of a sum of squared estimates divided by their respective variance. This sum is distributed asymptotically as χ^2 with 3+K+P degrees of freedom.

Extensions to hypotheses that consist of linear and nonlinear combinations of model parameters can be obtained in a straightforward way. In all cases, the Wald test necessitates the estimation of the full unrestricted model.

6.3.3. Likelihood Ratio Tests in Spatial Process Models

To obtain a Likelihood Ratio test on the significance of model parameters, two estimations are necessary: one with the coefficient in question included, one with that coefficient excluded (i.e., set to its constrained value of 0).

For example, a LR test on the significance of the autoregressive parameter ρ in the general model would be based on the log−likelihood in the full model, and the log−likelihood in a model without a spatially lagged dependent variable. Formally, as before in (6.15):

$$L(\theta) = - (n/2).\ln(\pi) + \ln |\Omega^{-1/2}.B.A| - (1/2) \ v'v$$

The restricted log−likelihood is:

$$L(\theta_R) = - (n/2).\ln(\pi) + \ln|\Omega_R^{-1/2}.B_R| - (1/2) \ \upsilon'\upsilon$$

with $\upsilon'\upsilon = (y - X\beta_R)'B_R'\Omega_R^{-1}.B_R.(y - X\beta_R)$.

The resulting LR test consists not only of the usual difference of sums of squared residuals ($\upsilon'\upsilon$ and $v'v$), but also of the difference between the Jacobian determinants:

$$L_R = (\upsilon'\upsilon - v'v) + 2 \ \{\ln|\Omega^{-1/2}BA| - \ln |\Omega_R^{-1/2}.B_R|\}$$

with the coefficients respectively evaluated at their restricted and unrestricted estimates. The resulting test statistic is asymptotically distributed as χ^2 with 1 degree of freedom, or, alternatively, its square root is distributed as a standard normal variate.

Similar expressions can be derived for the other parameters in the model, as well as for the simpler models that are encompassed in the general specification.

6.3.4. Lagrange Multiplier Tests in Spatial Process Models

In contrast to the Wald and Likelihood Ratio approaches, tests based on the Lagrange Multiplier principle do not necessitate the estimation of the more complex model. For the general specification considered here, this implies that ordinary least squares estimation will suffice for most spatial hypotheses of interest.

More specifically, consider a partitioning of the following form for the parameter vector in the general spatial model:

$$\theta' = [\rho \ \lambda \ \alpha' \mid \sigma^2 \ \beta']$$

where α has now been normalized so that the common error variance σ^2 can be isolated. The most general spatial null hypothesis would pertain to both spatial dependence (ρ and λ) as well as spatial heterogeneity (α), e.g.:

$$H_0 : [\rho \ \lambda \ \alpha'] = 0.$$

The LM test for this hypothesis would be:

$$LM = d'I^{11}d \sim \chi^2 \ (2+P)$$

In this expression, d is the score vector with elements (6.21)−(6.24), to be evaluated under the null hypothesis, i.e., with ρ, λ and α set to zero, and σ^2 and β set to their ordinary least squares estimates (with $\sigma^2 = e'e/N$ as the ML estimate). Indeed, under H_0, the general model reduces to a simple linear regression, with $A=B=I$ and $\Omega=\sigma^2 I$.

In many situations in non−spatial econometrics, the information matrix turns out to have a block−diagonal structure in the relevant coefficients, which greatly facilitates the derivation of the partitioned inverse I^{11}. The presence of a spatially lagged dependent variable, and the resulting covariance between the estimates for ρ and λ (and ρ and β) precludes this in the spatial case. As a consequence, the expression for the inverse I^{11} is more complex. [16]

As shown in Anselin (1988a), a formal derivation of the LM test for the null hypothesis of a joint presence of spatial dependence and spatial heterogeneity yields a complex statistic. It consists of two parts, one pertaining to the heteroskedastic component, the other to the spatially dependent component. The former results in the Breusch and Pagan (1979) statistic, in their notation:

$$(1/2).f'Z(Z'Z)^{-1}.Z'f \sim \chi^2 \ (P) \tag{6.35}$$

with

$$f_i = (\sigma^{-2}.e_i^{\ 2} - 1)$$

in which e_i is the least squares residual for observation i, and Z is the N by (P+1) matrix containing the z vectors for each observation.

The part of the test that pertains to spatial dependence does not have a direct counterpart in standard econometric results for serial autocorrelation and dynamic specifications. [17] Its presentation is greatly facilitated by the following simplifying notation:

$$R_y = e'W_1.y/\sigma^2$$

$$R_e = e'W_2.e/\sigma^2$$

$$M = I - X(X'X)^{-1}X'$$

$$T_{ij} = tr \{W_i.W_j + W_i'W_j\}$$

$$D = \sigma^{-2}.(W_1X\beta)'M.(W_1X\beta)$$

$$E = (D + T_{11}).T_{22} - (T_{12})^2$$

In this, R_y and R_e correspond to N times the regression coefficient of W_1 on y and W_2 on e respectively. The latter is the unstandardized Moran statistic for e. [18] The matrix M is the usual idempotent projection matrix from linear regression analysis. D is σ^{-2} times the residual sum of squares in a regression of $W_1.X\beta$ on X, i.e., a regression of the spatially lagged predicted values on the original regressors. The other expressions are introduced purely for notational sake.

The resulting LM test for spatial dependence has the form:

$$LM = E^{-1}.\{(R_y)^2.T_{22} - 2.R_y.R_e.T_{12} + (R_e)^2.(D+T_{11})\} \qquad (6.36)$$

and is asymptotically distributed as χ^2 with 2 degrees of freedom.

This rather awkward expression simplifies greatly for the case where the spatial weight matrices W_1 and W_2 are the same, and consequently,

$$T_{11}=T_{12}=T_{22}=T= tr \{(W' + W).W\}.$$

The resulting expression is:

$$D^{-1}.(R_y-R_e)^2 + (1/T)(R_e)^2 \qquad (6.37)$$

The second element in (6.37) is the LM test on spatial residual autocorrelation (distributed asymptotically as χ^2 (1)), reported in Burridge (1980).

The full test statistic for both spatial dependence and spatial heterogeneity is the sum of (6.36) and (6.35), or (6.37) and (6.35), asymptotically distributed as χ^2 with 2+P degrees of freedom. Since both the heteroskedastic and the residual spatial autoregressive elements in the LM statistic correspond to one–directional tests against a particular form of misspecification, one would suspect that the first term corresponds to a one–directional test for the omitted variable Wy. However, this is not the case. As shown in Anselin (1988a), the corresponding statistic, distributed as χ^2 with one degree of freedom, is:

$$(R_y)^2.(D+T)^{-1} \qquad (6.38)$$

The failure of the additivity of the one–directional tests to form the overall test is in contrast to the results in Jarque and Bera (1980) and Bera and Jarque (1982) for joint heteroskedasticity and serial correlation. This is due to the presence of the spatially lagged dependent variable, which results in a complex interaction between the estimates for ρ and λ. This interaction follows from the structural relationships between a spatial autoregressive process in the dependent variable and a spatial process in the disturbance, and will be considered again in the treatment of the common factor approach in Section 13.3.

The expressions (6.35)–(6.37) can form the basis for a wide range of tests, one–directional as well as multidirectional. None of these necessitates nonlinear estimation, and they can be fairly easily implemented in traditional regression packages that have some matrix manipulation capacities. A full illustration of the various combinations of tests that can be constructed in this manner is provided in the empirical applications of Section 12.2.

In addition, this general framework can be applied to several special cases, either by setting certain combinations of parameters to zero a priori, or by incorporating them in the estimation process. In the latter case, the estimation becomes more complex. Two special cases of particular interest to spatial analysis, dealing with spatial effects in the error terms of a linear regression model, are examined in Section 8.1.

6.3.5. Finite Sample Considerations

The Wald, Likelihood Ratio and Lagrange Multiplier tests are asymptotically equivalent. However, in finite samples they will typically result in different values, and it is not always clear how to interpret these apparent conflicts. [19] This issue has received substantial attention in econometrics, in the context of testing linear restrictions on the parameters of a linear regression model.

In general terms, the values for the test statistics conform to the following inequality in finite samples:

$$W \geq LR \geq LM.$$

Consequently, in small data sets, an uncritical use of the larger value for the Wald statistic will result in more frequent rejections of the null hypothesis, whereas the LM test will be more conservative. However, this inequality may be misleading, in the sense that it does not necessarily imply a difference in power between the tests (i.e., since the W tests results in more rejections, it would be more powerful), but rather a difference in size (i.e., the significance level, or probability of a type I error). When the proper size corrections are carried out, there is no uniformly more powerful test. [20]

Various corrections have been suggested in the context of the classical linear regression model, e.g., by Savin (1976), Berndt and Savin (1977), Breusch (1979), Evans and Savin (1982a), and Rothenberg (1982). Lagged dependent variables are included in Evans and Savin (1982b), and the case of a general non–spherical disturbance matrix is presented in Rothenberg (1984b). The corrections are typically in terms of the number of observations and degrees of freedom, and are mostly based on complex asymptotic approximations, such as the Edgeworth approximation. [21]

Although some of these results pertain to specifications which can also be applied to spatial modeling, such as the general nonscalar covariance case, many issues remain to be addressed before an extension of this finite sample approach can be achieved for the general spatial model.

In sum, considerable caution is needed when interpreting the indications given by asymptotic tests in finite samples. Additional insight in the sensitivity of the conclusions (at least in a qualitative sense) can be gained from some rather ad hoc *corrections*. One consists of using an *unbiased* estimate for the error variance instead of the more optimistic (smaller) ML−estimate. In other words, σ^2 is estimated as e'e divided by the degrees of freedom $(N-K)$ instead of by (the larger) N. Another approach would be to use the significance level indicated by an F−statistic (with q, $N-K$ degrees of freedom, with q as the number of constraints), for LM/q or W/q, instead of that associated with the asymptotic χ^2 (q) statistic. [22]

These approaches are fairly straightforward and easy to implement. However, the full implications for the relative power of the various tests have not been addressed, and remain to be investigated for situations of interest in spatial econometrics.

Appendix 6.A: **Some Useful Results on Matrix Calculus**

In this Appendix, I will present some useful elements of matrix calculus that are needed to derive the score vector and information matrix for the general spatial process model considered in this chapter.

For ease of exposition, I will frame the discussion in the context of the specific operations needed in Sections 6.2.4. and 6.2.5. More extensive surveys of matrix calculus can be found in Neudecker (1969), Rogers (1980), Magnus and Neudecker (1985, 1986), and in several econometrics texts, e.g., Theil (1971), Pollock (1979), Judge et al. (1985), and Amemiya (1985).

The log–likelihood function under consideration is, as in (6.15):

$$L = -(N/2).\ln(\pi) \; -(1/2).\ln|\Omega| \; +\ln|B| \; +\ln|A| \; -(1/2)v'v$$

with

$$v = \Omega^{-1/2}B.(Ay - X\beta)$$

For the derivation of the first partial derivative of this expression with respect to the elements of the parameter vector, a derivative of the natural log of a determinant is needed to deal with the partials of $|\Omega|$, $|A|$, and $|B|$ with respect to their parameters. Also, the rule for the partial derivative of a matrix product and of a quadratic form has to be applied several times to take the partial of the $v'v$ term with respect to each parameter. In the derivation of the information matrix, the partial derivative of the inverse of a matrix with respect to a scalar is needed.

The various rules are listed below, and illustrate the derivations needed to obtain the results in Section 6.2.

$$\partial(\rho W_1)/\partial\rho \qquad = W_1$$

$$\partial A/\partial\rho \qquad = \partial(I-\rho W_1)/\partial\rho$$

$$= \partial I/\partial\rho \; -\partial(\rho W_1)/\partial\rho$$

$$= -W_1$$

$$\partial\ln|A|/\partial\rho \qquad = \mathrm{tr}A^{-1}.\partial A/\partial\rho$$

$$= \mathrm{tr}A^{-1}.(-W_1)$$

$$\partial v/\partial\rho \qquad = \partial[\Omega^{-1/2}.B(Ay - X\beta)]/\partial\rho$$

$$= \Omega^{-1/2}.B.[\partial A/\partial\rho].y$$

$$= \Omega^{-1/2}.B.(-W_1).y$$

$$\partial v'v/\partial p \quad = v'(\partial v/\partial p) + (\partial v'/\partial p).v$$

$$= 2v'(\partial v/\partial p)$$

$$= 2.\Omega^{-1/2}.B.(-W_1).y$$

$$\partial A^{-1}/\partial p \quad = -A^{-1}.(\partial A/\partial p).A^{-1}$$

$$= -A^{-1}.(-W_1).A^{-1}$$

$$= A^{-1}.W_1.A^{-1}$$

$$\partial tr(A^{-1}W_1)/\partial p \quad = tr[\partial A^{-1}W_1/\partial p]$$

Appendix 6.B: **Derivation of the Elements of the Information Matrix for the General Model.**

The first step in the derivation of the elements of the information matrix for the general model discussed in this chapter consists of obtaining the second partial derivatives of the log–likelihood with respect to the elements of the parameter vector θ. A fairly straightforward but tedious application of the matrix calculus principles from Appendix 6.A. yields:

For the diagonal elements:

$$\partial^2 L/\partial\beta\partial\beta' = -X'B'\Omega^{-1}BX$$

$$\partial^2 L/\partial\rho^2 = -\text{tr}(A^{-1}W_1A^{-1}W_1) -(BW_1y)'\Omega^{-1}BW_1y$$

$$\partial^2 L/\partial\lambda^2 = -\text{tr}(B^{-1}W_2B^{-1}W_2) - [W_2(Ay-X\beta)]'\Omega^{-1}W_2(Ay-X\beta)$$

For the cross–product terms:

$$\partial^2 L/\partial\beta\partial\rho = \partial^2 L/\partial\rho\partial\beta'$$

$$= -(BX)'\Omega^{-1}BW_1y$$

$$\partial^2 L/\partial\beta\partial\lambda = \partial^2 L/\partial\lambda\partial\beta'$$

$$= -(BX)'\Omega^{-1}W_2(Ay-Xb) -\nu'\Omega^{-1/2}W_2X$$

$$\partial^2 L/\partial\beta\partial\alpha_p = \partial^2 L/\partial\alpha_p\partial\beta'$$

$$= - (1/2) (BX)'\Omega^{-2}H_pB(Ay-X\beta)$$

$$- (1/2)\nu'\Omega^{-3/2}H_pBX$$

$$\partial^2 L/\partial\rho\partial\lambda = \partial^2 L/\partial\lambda\partial\rho$$

$$= - [W_2(Ay-X\beta)]'\Omega^{-1}BW_1y -\nu'\Omega^{-1/2}W_2W_1y$$

$$\partial^2 L/\partial\rho\partial\alpha_p = \partial^2 L/\partial\alpha_p\partial\rho$$

$$= - (1/2)(BW_1y)'\Omega^{-2}H_pB(Ay-X\beta)$$

$$-(1/2)\nu'\Omega^{-3/2}H_pBW_1y$$

$$\partial^2 L/\partial\lambda\partial\alpha_p = \partial^2 L/\partial\alpha_p\partial\lambda$$

$$= - (1/2)[W_2(Ay-X\beta)]'\Omega^{-2}H_pB(Ay-X\beta)$$

$$- (1/2) \nu'\Omega^{-3/2}H_pW_2(Ay-X\beta)$$

$$\partial^2 L/\partial\alpha_p\partial\alpha_q \;\; = \;\; (1/2) \text{ tr } \Omega^{-2}H_pH_q - (1/2) \text{ tr } \Omega^{-1}H_{pq}$$

$$- (1/4)[B(Ay-X\beta)]'\Omega^{-3}H_qH_pB(Ay-X\beta)$$

$$- (3/4) \; v'\Omega^{-5/2}H_qH_pB(Ay-X\beta)$$

$$+(1/2) \; v'\Omega^{-3/2}H_{pq}B(Ay-X\beta)$$

To obtain the expected values, the following definitions and relations between the error terms are used:

$$\epsilon = Ay - X\beta$$

$$\mu = B(Ay - X\beta) = B\epsilon$$

$$v = \Omega^{-1/2}B(Ay - X\beta) = \Omega^{-1/2}\mu = \Omega^{-1/2}B\epsilon$$

It follows that, in terms of expected values:

$$E\,[\epsilon] = E\,[\mu] = E\,[v] = 0$$

$$E[\epsilon\epsilon'] = B^{-1}\Omega B'^{-1}$$

$$E[\mu\mu'] = \Omega$$

$$E[vv'] = I$$

and, for y:

$$y = A^{-1}X\beta + A^{-1}B^{-1}\Omega^{1/2}v = A^{-1}X\beta + A^{-1}B^{-1}\mu$$

$$E[y] \;\;\; = \;\;\; A^{-1}X\beta$$

$$E[yy'] \;\;\; = \;\;\; (A^{-1}X\beta)(A^{-1}X\beta)' + A^{-1}B^{-1}\Omega B'^{-1}A'^{-1}$$

An application of these properties to the above partial derivatives, in combination with a judicious use of the trace operator, yields the elements of the information matrix given in (6.25)–(6.34). As an illustration, consider $-E[\partial^2 L/\partial p^2]$. The first term is nonstochastic and poses no problems. In the second term, a trace operator can be introduced, since the expression, a quadratic form in y, is also a scalar. As a result:

$$E\,[(BW_1y)'\Omega^{-1}(BW_1y)] \;\;\; = E\,[\text{ tr }(BW_1y)'\Omega^{-1}(BW_1y)\;]$$

$$= E\,[\text{ tr }(W_1'B'\Omega^{-1}BW_1)(yy')\;]$$

$$= \text{ tr }\{(W_1'B'\Omega^{-1}BW_1).E[yy']\}$$

The final result follows in a straightforward manner by substituting the result for E[yy'] in this expression.

NOTES ON CHAPTER 6

[1]   As in the time series case, this is due to the complex stochastic nature of the inverse term. This will typically contain elements which are a function of the y's (and therefore the ϵ) at every observation point. As a result, this term will not be uncorrelated with ϵ. Moreover, while in time series $E[y_L{}'\epsilon]=0$ if there is no serial residual correlation, this is not the case in space. Indeed, in the spatial model: $E[y_L{}'\epsilon]=E\{[W.(I-\rho W)^{-1}\epsilon]'\epsilon\}$ which is only zero for $\rho=0$.

[2]   This in contrast to some of the early suggestions of Hordijk (1974). A more rigorous demonstration of this point is given in Anselin (1981).

[3]   As pointed out in the introduction to this chapter, these approaches are discussed in detail in Cliff and Ord (1981), Ripley (1981), and Upton and Fingleton (1985). See also Anselin (1980) for a detailed description of issues involved in the derivation of maximum likelihood estimators for spatial autoregressive models.

[4]   Previous formal treatments of this issue can be found in, e.g., Silvey (1961), Bar—Shalom (1971), Bhat (1974), and Crowder (1976), based on a conditional probability framework. However, it is only in the more recent articles that the simultaneous case is considered. This situation is most relevant to econometric models with lagged dependent variables and general error variance structures. See also the various laws of large numbers and central limit theorems discussed in Chapter 5.

[5]   A more rigorous formulation, for the situation where a normal distribution is assumed, is given in Heijmans and Magnus (1986c). The conditions presented there can be taken as the formal structure within which ML estimation can be carried out for the spatial autoregressive models considered here.

[6]   Here, as well as in the rest of the chapter, the notation det stands for determinant of a matrix. Where appropriate for notational simplicity, the symbol | | will be used as well.

[7]   The Jacobian in the first order spatial autoregressive model, $y = \rho W + \epsilon$, is det $(I-\rho W)$. Its properties have been explored in detail, and related to characteristics of the weight matrix on regular lattice structures, in Griffith (1980) and Ord (1981).

[8]   Given the usual specification of the model in terms of standardized weight matrices, the parameterization of A, B, and Ω will lead to well—behaved partial derivatives. However, for more general specifications, e.g., with a nonlinear distance decay function as spatial weights, the conditions needed to satisfy the existence of a continuously differentiable log—likelihood should also be checked.

[9]   It should be noted that the conditions (6.18)−(6.20) do not have to hold jointly to satisfy (6.17). They are sufficient, but too strict.

[10]   Ord (1975) has derived a simplification of determinants such as |A| and |B| in terms of their eigenvalues, more specifically as a product over i of $(1-\rho.\omega_i)$, where the ω_i are the eigenvalues of W. Consequently, the regularity conditions can be expressed in terms of these eigenvalues. Although this is usually taken as $-1 < \rho < +1$, in Anselin (1982, p.

1025) I showed that a more precise inequality is: $- (1/\omega_{max}) < \rho < +1$, where ω_{max} is the largest negative eigenvalue of W (in absolute value).

[11] In the notation below, tr stands for trace of a matrix, and α_p stands for the p-th element of the vector α, with p=0,1,...,P. H_p stands for the diagonal matrix with elements $\partial h/\partial \alpha_p$, where h is $h(z\alpha)$, or, explicitly, for $(\partial h/\partial s).z_p$, where $s=z\alpha$ and z_p is the p-th element of the z vector.

[12] Detailed results for the special models are given in Ord (1975), Anselin (1980), Cliff and Ord (1981), and Upton and Fingleton (1985).

[13] See in particular the literature on the estimation of random coefficient models, e.g., in Swamy (1971), Raj and Ullah (1981), and the overview in Amemiya (1985).

[14] In the literature, there is no uniform practice for estimating the asymptotic variance matrix for the parameters of spatial process models. The variance matrix is often estimated by its sample equivalent, and not necessarily based on the explicit derivation of the expected values of the second partial derivatives. Also, sometimes invalid simplifying assumptions have been introduced for the traces of the various matrices. For example, in Bivand (1984, p. 32), it is suggested for a mixed regressive spatial autoregressive model that $tr(W_1 A^{-1})^2$ equals $tr(W_1 A^{-1})'(W_1 A^{-1})$. Unless the weight matrices are symmetric (which they are not in Bivand's example), this will not hold, as demonstrated in Anselin (1988a), and further illustrated in the empirical applications in Section 12.2.

[15] See also the discussion of central limit theorems in Section 5.2., and the treatment of the properties of the ML estimator in the previous section.

[16] However, as shown in Anselin (1988a), under the null hypothesis, the block diagonality holds between the spatially dependent and heteroskedastic components.

[17] For an overview of the standard econometric approach, see e.g., Breusch and Godfrey (1981), Pagan and Hall (1983).

[18] See, e.g., Cliff and Ord (1981). This coefficient has also been called R_{01} in Anselin (1982) and Bivand (1984).

[19] This is in addition to the problem of the indeterminacy of the information matrix estimate associated with LM tests. In many situations this matrix can be estimated in a number of different ways, e.g., based on the expected values or on sample equivalents. For a review of this issue, see, e.g., Davidson and MacKinnon (1983).

[20] When only a single constraint is tested, appropriate size corrections result in equal power between the tests. For several constraints, the results are not clear. For more extensive discussion, see the references cited in the text.

[21] For overview, see, e.g., Sargan (1976), Phillips (1977, 1982), and Serfling (1980). A critical view on the relevance and usefulness of these procedures is expressed in a theoretical review by Taylor (1983), and in extensive Monte Carlo simulations by Kiviet (1985, 1986).

[22] This is based on the asymptotic convergence in distribution of an $F(q, N-K)$ statistic to a $\chi^2(q)/q$. Also, when the null hypothesis consists of only one constraint, the significance levels could be based on a more conservative Student–t variate (with $N-K$ degrees of freedom), instead of the standard normal.

CHAPTER 7

ALTERNATIVE APPROACHES TO INFERENCE

IN SPATIAL PROCESS MODELS

The maximum likelihood approach to estimation and hypothesis testing in spatial process models is by far the better known methodological framework. Moreover, in most of the literature in spatial econometrics it is the only technique considered and implemented. Nevertheless, several alternatives can be suggested to avoid some of the problems associated with ML estimation. Specifically, the numerical complexities of the nonlinear optimization and the restrictive parametric framework are features that the techniques discussed in this chapter attempt to deal with in a more satisfactory manner.

Three alternative approaches are considered in some detail: instrumental variables estimation, Bayesian techniques, and robust estimation methods. While these techniques have received widespread application in standard econometrics, their implementation in spatial models has been rather limited. In the discussion below, I will therefore stress those aspects of the various methods that need special consideration in a spatial context. Given the limited experience with these approaches in spatial econometrics, I will also outline a number of directions for research on issues that have not been fully resolved to date.

7.1. Instrumental Variables Estimation in Spatial Process Models

In many respects, the instrumental variables estimation method has similar asymptotic properties as the maximum likelihood approach, and is much easier to implement numerically. In spite of these attractive properties, its application to spatial models has been almost non−existent.

The discussion of the instrumental variable approach in the spatial economic literature is limited to a few references. It is considered as an estimating technique for a pure first order spatial autoregressive specification in Haining (1978a) and Bivand (1984). However, the technique is not actually implemented in either study. Moreover, the first order autoregressive specification for which it is discussed is of rather limited use in applied empirical work. [1]

In Anselin (1980, 1984a), a more general framework is outlined and illustrated empirically. It focuses on contexts within which instrumental variables estimation can be usefully implemented in models with spatially lagged dependent variables. The models considered are special cases of the general specification in Chapter 4, to the extent that they do not include potential heteroskedasticity.

In this section, I will discuss the ramifications of an instrumental variables (IV) approach to spatial modeling in some more detail. Since this type of technique is not treated in the well—known texts of Cliff and Ord (1981) and Upton and Fingleton (1985), it may be less familiar to spatial analysts. I will therefore elaborate on general methodological issues as well as outline specific spatial aspects of the IV technique.

The section consists of three parts. First, I outline the general principle behind the instrumental variables approach, and its properties and limitations within the context of models with spatially lagged dependent variables. Next, I consider some implementation issues for two forms of the mixed regressive spatial autoregressive model. One specification has regular error terms, while the other includes spatial dependence in the disturbance.

7.1.1. General Principles

As pointed out in Section 6.1.1, the failure of OLS in models with spatially lagged dependent variables is due to the correlation (in terms of expected values as well as asymptotically) between the spatial variable and the error term. In contrast to the time series case, this correlation occurs irrespective of the properties of the error term. This situation is similar to the estimation of parameters in a system of simultaneous equations, where the dependence between endogenous variables and error terms is at the root of the lack of consistency of OLS.

Intuitively, the instrumental variables approach is based on the existence of a set of instruments, Q, that are strongly correlated with the original variables $Z = [y_L, X]$, but asymptotically uncorrelated with the error term:

$$plim \ (1/N) \ Q'\epsilon = 0 \qquad (7.1)$$

and

$$plim \ (1/N) \ Q'Z = M_{QZ} \qquad (7.2)$$

a nonsingular finite matrix, with

$$y = [y_L, X]\theta + \epsilon = Z\theta + \epsilon \qquad (7.3)$$

Premultiplying both sides of equation (7.3) by $(1/N)$ times the instrument matrix Q', yields:

$$(1/N) \ Q'y = (1/N)Q'Z\theta + (1/N)Q'\epsilon$$

Using the properties of the *plims* that are assumed in $(7.1)-(7.2)$, this expression can be *solved* for the parameters θ, which results in the instrumental variable estimate θ_{IV}. Formally, since

$$plim \ (1/N)Q'y = plim \ (1/N)Q'Z\theta + 0$$

the estimate for θ follows as:

$$\theta_{IV} = [Q'Z]^{-1}Q'y \qquad (7.4)$$

provided that the matrix Q'Z is invertible. However, in most situations of interest, the number of instruments (q) will be larger than the number of parameters in the model, so that (7.4) is not well defined.

A more rigorous approach is based on the method of moments interpretation of instrumental variables estimation, in which it is viewed as a general optimization problem. [2] Specifically, the IV estimator is seen as the *solution* to the system of equations expressed in

$$Q'(y - Z\theta) = 0.$$

In the general case where the dimension of Q is larger than Z (i.e., more intruments than variables), no exact solution exists. Therefore, similar to the least squares approach to estimation, the problem can be formulated as a minimization of the quadratic distance from zero, in a norm S:

$$\min \phi(\theta) = (y - Z\theta)'Q.S.Q'(y - Z\theta)$$

A particular choice for S with attractive properties is $S = (Q'Q)^{-1}$, which yields the following minimization problem:

$$\min \phi(\theta) = (y - Z\theta)'Q.(Q'Q)^{-1}.Q'(y - Z\theta)$$

The solution to this optimization problem is the IV estimator, θ_{IV}:

$$\theta_{IV} = (Z'P_Q Z)^{-1}.Z'P_Q.y \tag{7.5}$$

with

$$P_Q = Q.(Q'Q)^{-1}.Q' \tag{7.6}$$

as an idempotent projection matrix. In the special case where Q is of the same column dimension as Z, the more familiar expression (7.4) results. The asymptotic covariance for the estimated parameters is:

$$\text{var}\ (\theta_{IV}) = \sigma^2 [M_{ZQ}.(M_{QQ})^{-1}.M_{QZ}]^{-1}$$

with as finite sample equivalent,

$$\text{var}\ (\theta_{IV}) = \sigma^2 [Z'P_Q Z]^{-1}$$

where the error variance is estimated as

$$\sigma^2 = (y - Z\theta_{IV})'(y - Z\theta_{IV})/N$$

The M are asymptotic nonsingular matrices of finite constants, defined in terms of the *plims* of the cross products of the instruments and the variables. Formally,

$$M_{ZQ} = plim\ (1/N)\ Z'Q$$

$$M_{QQ} = plim\ (1/N)\ Q'Q$$

$$M_{QZ} = plim \ (1/N) \ Q'Z$$

These conditions, together with $plim \ (1/N)Q'\epsilon = 0$, ensure consistency and asymptotic normality of the IV estimate, provided that $\{Z, \ Q, \ \epsilon\}$ is a proper mixing sequence. [3]

Computationally, expression (7.5) turns out to be equivalent to the more familiar two–stage least squares (2SLS) estimator for the parameters in an equation from a simultaneous system. Indeed, since the projection matrix P_Q is idempotent:

$$Z'P_QZ = Z'P_Q'P_QZ$$

Upon closer examination, P_QZ can be seen to correspond to a matrix of predicted values from regressions of each variable in Z on all the instruments in Q. Formally,

$$P_QZ = Q.\{(Q'Q)^{-1}QZ\}$$

with the term in brackets as the familiar OLS estimate for a regression of Z on Q. Therefore, with Z_p as the predicted values of the Z, the IV estimator (7.5) can also be expressed as:

$$h_{IV} = [Z_p'Z_p]^{-1}Z_p'y$$

or, equivalently, due to the idempotency of the projection matrix, as:

$$h_{IV} = [Z_p'Z]^{-1}Z_p'y$$

This is the familiar 2SLS estimator. It is equivalent to OLS on the predicted values for the explanatory variables obtained from an auxiliary regression on a fixed set of *exogenous* instruments.

7.1.2. Implementation Issues in Models with Spatially Lagged Dependent Variables

The approach outlined in the previous section can be implemented for models with a spatially lagged dependent variable in a straightforward manner, provided that a proper set of instruments can be found. Moreover, estimation can be carried out by means of standard regression packages, instead of having to resort to more specialized numerical optimization routines needed for ML.

Although in the narrow sense used here, only the spatially lagged variable is considered as a cause of problems, the IV approach is applicable to a more general formulation as well. In particular, it can easily be extended to situations where some of the other explanatory variables may be stochastic or endogenous variables in a system of equations.

The properties of the IV estimator are only asymptotic. Although consistent, it is generally not the most efficient estimator. The efficiency depends in a crucial way upon the *proper* choice of the instruments. Also, the asymptotic normality does not carry over into a well defined finite sample distribution, so that inference in realistic data sets may be problematic. [4]

In addition, the selection of the instruments is a major implementation problem. In principle this should be based on a theoretical framework. However, in practice there is little formal guidance and the selection of the instruments is likely to be largely ad hoc. [5]

Ideally, the instruments should be strongly related to the original variables, and asymptotically uncorrelated with the error terms. Although there is no general rule to achieve this, the first requirement can be checked fairly easily, e.g., by means of a canonical correlation approach. [6] The choice of the number of instruments can be guided by similar techniques. While asymptotically the number of instruments does not matter, in finite samples practical limits will result from problems with multicolinearity and concerns about degrees of freedom.

The requirement of asymptotic uncorrelatedness between instruments and the error term is much harder to assess. In some simple cases where the instruments consist of fitted values, the lack of correlation can be shown analytically. When the problem is more complex, several specification tests can be used to assess the *exogeneity* of the instruments, such as the Wu–Hausman approach. Since these tests can be applied in a straightforward way to spatial models, they will not be further discussed here. [7]

In practice, few specific suggestions have been offered for the choice of instruments for the spatially lagged variable. Since the other variables (the X) can presumably act as instruments for themselves, attention typically focuses only on the spatial variables.

Haining (1978a) suggested, in the context of a first order spatial autoregressive model, to take a set of observations on the dependent variable from a previous time period. Since this would imply the availability of a time–space data set, it would not be very practical nor an efficient use of data. Indeed, in this situation a purely cross–sectional approach would clearly result in a loss of information. In Anselin (1980), a number of suggestions are made for the general spatial autoregressive model. For example, the use of a spatially lagged predicted value of a regression of the dependent value on the non–spatial regressors is shown to be an acceptable instrument. [8] Alternatively, spatial lags of the exogenous variables in the model can be used, although this may lead to problems with multicolinearity. In general, any combination of exogenous variables, spatially lagged or not, is likely to give satisfactory results. [9]

The various exogeneity tests can also be used to assess the general seriousness of the problem caused by the presence of the spatially lagged dependent variable. In some situations, this effect may not warrant the use of an IV approach (or an ML approach). Although consistent, in small samples the IV estimate may be inferior to OLS in terms of mean squared error. A trade–off between the potentially smaller MSE of the biased OLS estimate and the consistency of the IV estimate is obtained in the so–called k–class or shrinkage estimators. In general terms, these are constructed from a weighted average between the two estimators, where the weights are obtained in some *optimal* fashion. [10] In some instances these estimators may result in superior performance. The application of this approach in a spatial context should be straightforward, although its properties in this situation remain to be investigated.

A final issue of practical importance in the use of IV estimators for spatial models is the potential for results that imply an explosive pattern of spatial dependence. Indeed, the estimate for the spatially autoregressive parameter is unrestricted and may be larger than one, which would indicate a lack of stability (for a standardized weight matrix). The same result can occur in OLS estimation. Typically, it is avoided in the numerical optimization used to obtain ML estimates, by forcing the stability constraint to hold in every iteration. The large parameter value may be an indicator of a poor choice of instruments (which led to an inefficient estimate) or may point to a general misspecification of the model.

7.1.3. Implementation Issues in Models with Spatially Lagged Dependent Variables and Spatially Autoregressive Error Terms

The model with spatial dependence in both the dependent variable and the error term is similar to the specification in time series analysis (and distributed lags) for which the use of the IV estimator was first advocated, e.g., by Liviatan (1963) and Wallis (1967). [11] However, the extent to which various procedures developed for the time series context can be applied to the spatial model is rather limited.

For example, consider the specification:

$$y_t = \rho y_{t-1} + X_t \beta + \epsilon_t \tag{7.7}$$

$$\epsilon_t = \lambda \epsilon_{t-1} + \mu_t \tag{7.8}$$

which is formally equivalent to the spatial model:

$$y = \rho W_1 y + X\beta + \epsilon \tag{7.9}$$

$$\epsilon = \lambda W_2 \epsilon + \mu \tag{7.10}$$

The standard IV approach to (7.7)−(7.8) takes into account the nonspherical nature of the disturbance in the adjusted estimator: [12]

$$\theta_{IVG} = [Z'Q(Q'\Omega Q)^{-1}Q'Z]^{-1}.Z'Q(Q'\Omega Q)^{-1}Q'y \tag{7.11}$$

with as estimate for the coefficient variance:

$$\text{var } \theta_{IVG} = [Z'Q(Q'\Omega Q)^{-1}Q'Z]^{-1} \tag{7.12}$$

To implement this estimator in practice, it is necessary to obtain a consistent estimate for the error covariance matrix Ω, or alternatively, for the parameters on which Ω depends. In the time series case, this is based on two−step or iterative approaches. The parameters for the error structure, e.g., the λ in model (7.8), are estimated by OLS from the residuals in a first−stage IV estimation for the model, as in the three pass least squares approach of Wallis. Since OLS is a consistent estimator for λ in model (7.8) as long as the errors μ are not autocorrelated, this approach has an intuitive appeal. However, in the spatial model, these properties no longer hold, as pointed out in Section 6.1. Moreover, it is not clear that an IV approach on the residuals would necessarily yield a consistent estimate.

Consequently, the iterative IV approach based on the residuals from a first stage may not be very efficient for models (7.9)−(7.10). An alternative consists of using the nonlinear form of this specification, which has the classical error terms. Formally, as in Section 6.2,

$$\epsilon = (I - \lambda W_2)^{-1}\mu$$

and thus

$$y = \rho W_1 y + X\beta + (I - \lambda W_2)^{-1}\mu$$

$$(I - \lambda W_2)y = (I - \lambda W_2)\rho W_1 y + (I - \lambda W_2)X\beta + \mu$$

or

$$y = \rho W_1 y + \lambda W_2 y - \lambda\rho W_2 W_1 y + X\beta - \lambda W_2 X\beta + \mu \qquad (7.13)$$

or

$$y = \alpha_1 W_1 y + \alpha_2 W_2 y - \alpha_3 W_2 W_1 y + X\beta_1 - W_2 X\beta_2 + \mu \qquad (7.14)$$

with the following constraints on the coefficients:

$$\alpha_1 . \alpha_2 = \alpha_3$$

$$\beta_2 . / \beta_1 = -\alpha_2$$

where ./ stands for element by element division, and thus the second equation implies K constraints.

The parameters in model (7.13) can be estimated by linear or non−linear IV. In the linear approach, the most practical strategy is to ignore the constraints on the parameters in the estimation. These nonlinear constraints can then be tested using an IV−analog of the Wald or LM tests (i.e., using the IV variance estimate as the equivalent of the inverse information matrix). Rejection of the constraints would point to a general failure of the model specification.

Alternatively, the linear approach could be taken as a first stage to obtain a consistent estimate for λ, to be used in a second iteration. The second iteration would be based on the standard estimator (7.11) with the λ used to estimate the error covariance Ω. [13] However, expression (7.13) potentially suffers from an identification problem when the two weight matrices are the same.

In that situation, the model becomes:

$$y = (\rho+\lambda)Wy - \lambda\rho W^2 y + X\beta - \lambda WX\beta + \mu$$

or

$$y = \alpha_1 Wy - \alpha_2 W^2 y + X\beta_1 - WX\beta_2 + \mu$$

from which two different estimates of ρ can be obtained. [14] In this situation, it is more practical to carry out the two–step approach mentioned above.

The nonlinear approach would consist of minimizing, by numerical techniques, the quadratic form:

$$g(Z,\theta)'Q(Q'Q)^{-1}Q'g(Z,\theta)$$

with

$$g(Z,\theta) = y - \rho W_1 y - \lambda W_2 y + \lambda\rho W_2 W_1 y - X\beta + \lambda W_2 X\beta$$

$$\theta' = [\rho \ \lambda \ \beta']$$

In this specification as well, there will be an identification problem when the two weight matrices are the same.

Clearly, the analysis of the spatial model is more complex than its time series counterpart. Although the instrumental variables approach provides a useful estimation framework, its properties in finite samples of interest in applied regional science are largely unknown and remain to be investigated.

In sum, the main attractiveness of the IV approach consists of its numerical simplicity and more direct integration within the main body of econometric theory. It also provides a easy way to introduce robust estimation into spatial econometrics, as outlined in Section 7.3.

Its main drawbacks are the potential of estimating explosive spatial autoregressive coefficients, implementation problems with the choice of the instruments, and the lack of well–understood finite sample properties.

7.2. Bayesian Approaches to the Estimation of Spatial Process Models

In a Bayesian approach to spatial modeling, the data analysis is viewed as part of a formal decision process that combines the construction of hypotheses, estimation of parameters and choice between models. The main philosophical difference with a traditional approach lies in the subjectivist view of probability. This leads to the integration of prior assumptions about the model and its parameters into a sequential estimation process, which allows a more formal treatment of issues of robustness and finite sample properties.

The range of results developed in the standard Bayesian literature in statistics and econometrics is extensive. [15] This has not been reflected in many relevant applications in regional science and geography. Although issues such as the choice of the spatial weight matrix, spatial forecasting and interpolation seem to be prime candidates for a Bayesian approach, the analytical and numerical problems associated with the implementation of these techniques in realistic contexts are severe. This is a likely cause for the lack of diffusion of these methods to applied empirical work in regional science. [16]

In line with this, the application of Bayesian techniques to the estimation of spatial process models has been of limited scope. Some fairly straightforward extensions of results from times series analysis to first order spatial autoregressive models and to regression models with spatially autocorrelated errors were suggested by Hepple (1979) and Anselin (1980, 1982). They are briefly outlined below. The Bayesian approach to model validation is considered in more detail in Section 14.3.

7.2.1. A Bayesian Approach to Inference in a First Order Autoregressive Model

Statistical inference in the Bayesian framework is based on a combination of prior information about the distribution of the parameters of the model with the information contained in the data set. Typically, when no precise assumptions are available, the prior distribution is taken to be diffuse or non−informative. The information in the data set is represented by the likelihood function.

The combination of prior distribution and likelihood is carried out by means of Bayes' law of probability. Specifically, the information contained in the model can be expressed as a joint distribution or density function in the variables and parameters, $h(\theta,y)$. In terms of conditional probabilities, this is equivalent to:

$$h(\theta,y) = f(y|\theta).g(\theta) = g(\theta|y).f(y)$$

where f pertains to the distribution of the data and g to the distribution of the parameters. In the Bayesian terminology, $g(\theta|y)$ is the posterior density, i.e. the information about the parameters after the data have been observed. Similarly, $g(\theta)$ is the prior density, i.e., the information (assumptions, prior convictions) about the parameters before the data were observed. Bayes' Theorem provides a formal way to find one from the other, via the observation of data, expressed in the likelihood function, $f(y|\theta)$. Formally,

$$g(\theta|y) = f(y|\theta).g(\theta)/f(y).$$

In order to operationalize this approach, it is necessary to define the prior densities for the model parameters and to assume an underlying distribution that leads to a well defined likelihood. In the first order spatial autoregressive model (with y as deviations from the mean):

$$y = \rho W y + \epsilon$$

the usual approach is to assume an underlying normal distribution for the error term. Consequently, the likelihood function is, as before (ignoring constants):

$$L \propto |I-\rho W|.\sigma^{-N}.\exp\{-(2\sigma^2)^{-1}.y'(I-\rho W)'(I-\rho W)y\} \tag{7.15}$$

The two parameters in the model are the autoregressive coefficient ρ and the error variance σ^2. Following the standard approach in econometrics, diffuse prior densities for these parameters are expressed as:

$$P(\sigma) \propto \sigma^{-1} \qquad\qquad 0 < \sigma < +\infty$$

$$P(\rho) \propto \text{constant} \qquad\qquad -1 < \rho < +1$$

Assuming σ and ρ to be independent results in a joint prior density for the parameters of the form:

$$P(\rho,\sigma) \propto \sigma^{-1} \tag{7.16}$$

The joint posterior distribution for the model parameters is found from a direct application of Bayes' law, multiplying (7.16) with (7.15):

$$P(\rho,\sigma|y) \propto |I-\rho W|.\sigma^{-(N+1)}.\exp\{-(2\sigma^2)^{-1}y'(I-\rho W)'(I-\rho W)y\} \tag{7.17}$$

Inference about the parameter ρ can be based on the properties of expression (7.17), after integrating out the nuisance parameter σ. The resulting marginal posterior distribution is of the form: [17]

$$P(\rho|y) \propto |I-\rho W|.\{y'(I-\rho W)'(I-\rho W)y\}^{-N/2} \tag{7.18}$$

The normalizing constant which ensures that (7.18) is a proper density function can be found by integrating (7.18) over the range -1 to $+1$ for ρ and setting the result equal to 1. Once this constant is obtained, the various moments of this posterior probability can be derived, as well as the exact probability of the parameter being in a particular range. For example, with a normalizing constant of Q, the mean of the posterior distribution can be found as:

$$\rho_p = Q \;_{-1}\!\int^{+1} \rho.|I-\rho W|.\{y'(I-\rho W)'(I-\rho W)y\}^{-N/2} \, d\rho.$$

For a quadratic loss function, this mean of the posterior density can be shown to be the optimal point estimator or minimum expected loss estimator (MELO). In large samples, the likelihood part of the posterior density will swamp the influence of the diffuse prior and the MELO estimate will tend to the ML estimate. However, in small samples these estimates will not be the same. In contrast to the asymptotic ML, the Bayesian approach allows for a precise assessment of the finite sample probability. [18]

7.2.2. A Bayesian Approach to Inference in the Linear Model with Spatially Autoregressive Error Terms

The same general principle as used in the previous section can be applied to the linear regression model with spatially autocorrelated errors. The approach is an extension of the time series case developed by Zellner and Tiao (1964), but differs in two respects: the Jacobian $|I-\lambda W|$ is more complex in the spatial case; and, no initial value needs to be specified in the spatial case, although the issue of starting values becomes more complex and leads to a general boundary value problem.

As before, diffuse prior densities are assumed for the model parameters:

$$P(\lambda) \propto \text{constant} \qquad\qquad -1 < \lambda < +1$$

$P(\beta) \propto$ constant $\qquad\qquad |\beta| < \infty$

$P(\sigma) \propto \sigma^{-1} \qquad\qquad 0 < \sigma < +\infty$

which, combined with an assumption of independence, yields the prior density as:

$P(\lambda,\sigma,\beta) \propto \sigma^{-1}.$

The likelihood for the model is of the form (ignoring constants):

$L \propto |I - \lambda W|.\sigma^{-N}.\exp\{-(1/2\sigma^2)(y - X\beta)'(I - \lambda W)'(I - \lambda W)(y - X\beta)\}.$

Consequently, a straightforward application of Bayes' theorem yields the posterior density for the model parameters as:

$$P(\lambda,\sigma,\beta) \propto |I - \lambda W|.\sigma^{-N}.\exp\{-(1/2\sigma^2)(y - X\beta)'(I - \lambda W)'(I - \lambda W)(y - X\beta)\} \qquad (7.19)$$

Typically, the error standard deviation σ is considered as a nuisance parameter and is integrated out, yielding a joint posterior density for λ and β: [19]

$$P(\lambda,\beta) \propto |I - \lambda W|.\{(y - X\beta)'(I - \lambda W)'(I - \lambda W)(y - X\beta)\}^{-N/2} \qquad (7.20)$$

Expression (7.20) forms the starting point for the derivation of various marginal and joint posterior distributions for combinations of the parameters. Similar to the approach in the previous section, normalization constants can be found from integration over the interval of allowable values for the coefficients to ensure that the results are proper density functions. In practice, the various integrations need to be based on numerical techniques, which limits the applicability of the Bayesian approach to rather simple models. However, these constraints on the scope of numerical integration are likely to become irrelevant as computer technology evolves.

As pointed out in the introduction to this section, the Bayesian approach has not yet been applied to many estimation issues in spatial econometrics. However, it would seem that the rigorous integration of the various steps in the modeling process into an overall decision framework would lead to a powerful basis to deal with several *tricky* problems in spatial modeling. In particular, the serious methodological problems associated with the prior assumptions involved in the choice of a spatial weight matrix seem to be most fruitfully approached from a Bayesian perspective. This still remains an area in which considerable research needs to be carried out.

7.3. Robust Approaches to Spatial Process Models

In contrast to the techniques discussed in the previous two sections, robust approaches have been well represented in spatial analysis. Nonparametric tests for spatial autocorrelation, and the interpretation of test statistics based on randomization and permutation techniques make up a substantial part of the relevant literature. A familiar example is the interpretation of the Moran test for spatial autocorrelation by Cliff and Ord (1973), based on the permutation principle of Hope (1968), and Edgington (1969, 1980). The nonparametric approach is also exemplified in the spatial tests developed by Sen and Soot (1977), Tjostheim (1978), Glick (1982), and in the work of Hubert and Golledge. [20]

More recently, there has been an expression of increased awareness of the relevance of robust approaches for empirical analysis in regional science and geography, e.g., in Costanzo (1983) and Knudsen (1987). [21] This is also reflected in the application of newly developed statistical techniques such as the bootstrap and jackknife to spatial modeling situations, e.g., by Stetzer (1982a), Folmer and Fischer (1984), and Folmer (1986).

In this section, I will briefly discuss the application of one particular robust technique, the bootstrap, to statistical inference in spatial process models. Although this method is increasingly well known, a consideration of the implications of spatial dependence (and to a lesser extent, spatial heterogeneity) for its implementation has not received much attention. After a short overview of the technique, I focus on this aspect in more detail below.

In addition to the bootstrap (and the related jackknife), a great many other robust techniques have been developed and suggested for use in econometrics. [22] Most methods are aimed at lessening the impact of one or more sources of potential misspecification, pertaining to the distributional assumptions (normal distribution) or to the functional form. Examples of fairly recent techniques are bounded influence estimation, and pseudo- and quasi-maximum likelihood estimation. [23] So far, these techniques have not seen much application in spatial analysis, which may be due in part to their predominant reliance on assumptions of independence. The extension and scope for application of this growing body of robust estimation methods to situations with spatial dependence and spatial heterogeneity largely remains to be investigated.

7.3.1. Bootstrapping in Regression Models: General Principles

The bootstrap and the related jackknife are examples of resampling techniques, which have recently received increased attention in statistics and econometrics. The principle behind these techniques is to use the randomness present in artificially created *resampled* data sets as the basis for statistical inference. This leads to alternative parameter estimates, measures of bias and variance, and the construction of pseudo significance levels and confidence intervals. [24]

In the case of the bootstrap, a series of estimates are calculated for a large number of data sets, obtained by random sampling (with replacement) from the original observations. Consequently, the same observation can be included more than once in the pseudo data, or not at all. Typically, the underlying sampling scheme assigns equal probability to each observation (i.e., $1/N$ if there are N observations). As a result, the pseudo data are taken to be independently and identically distributed (i.i.d). More complex approaches are possible as well, e.g., by taking a mixture of the i.i.d. assumption and a particular density function such as the normal (i.e., a smoothed bootstrap).

The statistic of interest is calculated for each replication of the pseudo data, and its empirical frequency distribution is used to derive estimates of parameters, bias and variance. For example, take R replications of random sampling from a given data set x, which results in pseudo−data sets x_1, x_2,..., x_R. For each of these, an estimate for the parameter of interest $\theta = f(x)$ is obtained (i.e., as a statistic, considered as a function of the observations), as θ_1, θ_2,..., θ_R. The bootstrap estimate for the parameter would be:

$$\theta_B = (1/R) \, \Sigma_r \, \theta_r$$

i.e., the mean of the estimates over all replications. The associated variance is measured as: [25]

$$\text{var}[\theta_B] = [1/(R{-}1)].\, \Sigma_r \, (\theta_r - \theta_B)^2$$

The difference between the bootstrap estimate (from all replications) and the parameter estimate derived from the original data set gives a measure of the bias of the estimate.

In regression analysis, there are two approaches that lead to a bootstrap estimate, one based on residuals, the other on observation points in multidimensional space. In the first approach, the resampling is based on a set of regression residuals, obtained from a first−step estimation, typically by OLS. In formal terms, for the population model:

$$y = X\beta + \epsilon$$

the OLS residuals are:

$$e = y - Xb.$$

with $b = (X'X)^{-1}X'y$, as a vector of OLS estimates.

An empirical distribution function is assumed for the residuals, which typically assigns equal probability to each. In other words, each residual will have probability $1/N$ (for N observations) for being drawn in the resampling scheme. Although OLS residuals are well known not to be independent (even for independent underlying errors), they are assumed to be so for the purposes of the resampling scheme. An alternative would be to use *independent* residuals such as BLUS residuals, although this is seldom done in practical implementations of the regression bootstrap. Another transformation of the residuals consists of *inflating* them to correct for the degree of underestimation as a result of the model fit. In the standard case, a factor of $[N.(N{-}K)]^{1/2}$ is often suggested. For more complex models there is no uniform approach. [26]

In general terms, a bootstrap replication is constructed from a set of randomly sampled residuals (with replacement, so that the same residual can occur more than once), in combination with the first−stage parameter estimates:

$$y_r = e_r + Xb$$

where e_r is a vector of resampled residuals, b is the first—step estimate and X is the matrix of observations on the fixed (exogenous) variables. An estimate for β from this pseudo data set is obtained by the same method as for the initial complete sample. For example, for OLS, this would yield:

$$b_r = (X'X)^{-1}X'y_r$$

This process is replicated a large number of times (e.g., R) to generate an empirical frequency distribution of the b_r. The bootstrap estimate for β is taken as the mean in this empirical distribution:

$$\beta_B = (1/R). \Sigma_r\, b_r \tag{7.21}$$

and its associated variance matrix is:

$$\text{var } (\beta_B) = [1/(R-1)].\Sigma_r\, (b_r - \beta_B)(b_r - \beta_B)' \tag{7.22}$$

The alternative approach is to generate pseudo data by random sampling (with replacement) from the N points in K+1 dimensional space represented by the vectors $[y_i\ x_i]$. Here, y_i is the i—th observation on the dependent variable, and x_i is a K—dimensional row vector corresponding to the i—th observation for the explanatory variables. Again, for each resampled pseudo data set an estimate b_r is obtained, and the full set of all b_r forms the basis for the bootstrap estimate β_B and its variance, as in (7.21)−(7.22).

Although the two approaches are asymptotically equivalent, in finite samples they typically yield different results. [27]

The main contribution of the bootstrap approach in regression analysis is to provide an alternative variance estimate in finite samples, for situations where the distributional assumptions for the error terms may not be satisfied. There is some evidence that the more conservative bootstrap estimate has a higher degree of realism than the variance yielded by the finite sample implementation of asymptotic variance matrices. [28] A similar role can be played by the jackknife. This is considered in more detail in the context of robust residual variance estimation, in Section 8.3, and illustrated empirically in Section 12.2.6.

7.3.2. Bootstrapping in Models with Spatial Dependence

In models with spatial dependence, and particularly with spatially lagged dependent variables, two issues need special consideration in the implementation of a bootstrap approach. One pertains to the estimator that should be used in the initial stage, as well as for each resampled pseudo data set. The other issue is more fundamental, and relates to the design of the resampling itself. I consider this first.

The point of departure for the construction of pseudo data is the independence of observation vectors or error terms. This is necessary to ensure that an empirical density can be assumed with equal probability for each observation point. Also, it prevents the random resampling from destroying inherent structural characteristics of the data set. When spatial dependence is present, the data points are obviously not independent.

As a consequence, a random sampling from N observation vectors $[y_i, (Wy)_i, x_i]$ is not very meaningful. An alternative is to take the residual approach, in analogy to the bootstrap in systems of simultaneous equations.

The spatial process models considered in this and the previous chapter can be reformulated as nonlinear models with independent errors. This is clearly the case for the simple mixed regressive autoregressive specification, but also for the model with spatially dependent error terms. Consequently, a resampling strategy needs to be formulated for models of the form:

$$y = f(y,X,\theta) + \mu$$

where μ is a vector of independent error terms. More specifically, consider the mixed regressive autoregressive model:

$$y = \rho Wy + X\beta + \mu$$

An estimate for the error vector can be obtained from the residuals u that result from the usual estimates of ρ and β, say r and b:

$$u = y - rWy - Xb$$

Similar to the bootstrap approach in the more traditional regression context, pseudo error terms can be generated by random sampling from vector u. For each of these, a pseudo vector of dependent variables can be found as:

$$y_r = (I - rW)^{-1}(Xb + u_r)$$

with X as the fixed (exogenous) variables, and b and r as the first round estimates. An estimate r_r and b_r can be obtained from applying an appropriate estimator to the regression of y_r on Wy_r and X. The bootstrap estimate for the model coefficients and their associated variance can then be obtained in the usual fashion, as in $(7.21)-(7.22)$. The random assignment of the (assumedly independent) error terms ensures that the spatial structure of the data is maintained. This would not be the case in the other approach.

A second implementation issue in spatial models relates to the choice of the estimator for the coefficients in the model. As pointed out in the previous chapter, OLS is clearly unsatisfactory. An alternative would be maximum likelihood. However, it would be computationally demanding, since the associated nonlinear optimization would have to be carried out for each of the many resampled data sets. Although not ideal, due to the problem with the choice of instruments, IV estimation remains as the most practical option. Clearly, these issues have not received much attention in spatial econometrics, and much still needs to be investigated.

A common attractive feature of the various resampling techniques is their heavy emphasis on computation, which has only become operational through the availability of the vastly increased capabilities of the newer computers. To some extent, the refined analytics and restrictive assumptions of the more traditional approaches are replaced by raw computation. As a result, some of the constraints imposed by the non−experimental nature of empirical research of human behavior may become less binding, although there is clearly a trade−off between robustness and specificity. The implications of these issues for spatial analysis remain to be more fully explored.

NOTES ON CHAPTER 7

[1] Hordijk and Nijkamp (1977, 1978) suggest the use of instrumental variables estimators in a space—time context. However, their actual implementation of this technique pertains to the time domain only. Also in Fisher (1971) this method is alluded to, although not implemented in a spatial application.

[2] For an extensive and rigorous treatment of these issues, see, e.g., Sargan (1958), Gallant and Jorgenson (1979), Hansen (1982), Bowden and Turkington (1984), and White (1984).

[3] See White (1984) for details on the formal conditions for asymptotic normality in a wide range of situations.

[4] This issue has received considerable attention in econometrics, in the context of 2SLS estimators for systems of simultaneous equations. Although the literature is substantial, the issues involved are complex and there is no clear and generally accepted solution. Overviews are given in, e.g., Phillips (1980), Mariano (1982), and Magdalinos (1985).

[5] For example, in Bowden and Turkington (1984, p. 85), the discussion of the proper selection of instruments is concluded by: "...the choice of effective instruments is best approached in a rather ad hoc manner, using what is known about the structural properties of the particular model being fitted."

[6] For a detailed discussion of the choice of minimal instruments and the role of canonical correlations in this respect, see, Bowden and Turkington (1984).

[7] The reader is referred to the standard econometric literature, e.g., Wu (1973, 1974), Hausman (1978, 1983), Hwang (1981), Spencer and Berk (1981), Holly (1982), and Ruud (1984).

[8] Formally, the instrument is Wy_p, with $y_p = X(X'X)^{-1}X'y$.

[9] This approach is illustrated in Anselin (1984a), for a spatial autoregressive model of housing values. For further empirical illustrations, see also the results in Section 12.3.3.

[10] See Theil (1971), and Bowden and Turkington (1984), for more details.

[11] See also Hendry and Srba (1977) for a more recent assessment.

[12] The estimator in (7.11)−(7.12) is the so−called OLS analog for this model. There is also a GLS analog of this IV estimator, in which every cross product is weighted by Ω^{-1}. See Bowden and Turkington (1984, Chapter 3) for full details. Since the distinction between the two approaches is not relevant to the spatial model, it is not further considered here.

[13] See also the treatment of spatial errors in regression models in Sections 8.2.2. and 13.3, for a similar approach in the spatial Durbin method.

[14] This is not a problem unique to the IV approach, but a result of the model specification. The same issue complicates the numerical optimization for the general model from Chapter 4. For a more extensive discussion and an empirical illustration, see Anselin (1980, Chapter 6).

[15] For an overview, see, e.g., Box and Tiao (1973), Fienberg and Zellner (1975), and the various writings of Zellner, in particular Zellner (1971, 1980, 1984, 1985).

[16] Some applications to spatial interaction analysis and urban modeling are discussed in March and Batty (1975), and Odland (1978).

[17] The approach is based on an extension of the examples given in Zellner (1971). For a detailed derivation, see Anselin (1982, pp. 1025-6).

[18] An extensive comparison of the small sample properties of the Bayesian MELO estimator with other techniques, based on Monte Carlo simulation, is given in Anselin (1980, 1982).

[19] The detailed derivation is based on the properties of the inverted gamma-2 density and can be found in Anselin (1980, Chapter 5).

[20] See, e.g., Hubert, Golledge and Costanzo (1981), Hubert and Golledge (1982a), Hubert, Golledge, Costanzo and Gale (1985), and Hubert (1985).

[21] Earlier opposition of an uncritical application of parametric techniques to issues in geography and regional science was voiced in Gould (1970, 1981).

[22] For a review, see, e.g., Huber (1972, 1981), Mosteller and Tukey (1977), and Koenker (1982).

[23] For more examples, see the review by Koenker (1982). Bounded influence estimation is presented in Belsey, Kuh and Welsch (1980), Welsch (1980), Krasker (1981), Krasker and Welsch (1982, 1985). The background on pseudo maximum likelihood can be found in White (1982), and Gourieroux, Monfort, and Trognon (1984a, 1984b). An application to spatial interaction modeling is given in Baxter (1985).

[24] For extensive overviews of the principles and formal properties of the bootstrap and jackknife, see, e.g., Miller (1974), Efron (1979a, 1979b), Efron (1982), Efron and Gong (1983). Specific issues related to the application in regression analysis and econometrics are discussed in Freedman (1981), Bickel and Freedman (1983), and Freedman and Peters (1984a, 1984b).

[25] For a more rigorous treatment, see Efron (1982) in particular.

[26] See Freedman and Peters (1984a, 1984b) for details.

[27] In the standard case (i.i.d. errors), the residual-based approach yields a variance matrix which is equivalent to the usual OLS result, while the resampling based on observation vectors does not. However, there is some evidence that the latter is more robust to functional misspecification. See Efron (1982), and Efron and Gong (1983), for further discussion.

[28] Freedman and Peters (1984a) find this to be the case for a SUR model. However, these results were not replicated in a simultaneous equation context in Freedman and Peters (1984b).

CHAPTER 8

SPATIAL DEPENDENCE IN REGRESSION ERROR TERMS

The analysis of the effects of spatial dependence in the error terms of the linear regression model was the first specifically *spatial* econometric issue to be addressed in the regional science literature. Initial problem descriptions and the suggestion of some solutions were formulated in the early 1970's, e.g., by Fisher (1971), Berry (1971), Cliff and Ord (1972), McCamley (1973), Hordijk (1974), Martin (1974), Bodson and Peeters (1975), and Hordijk and Paelinck (1976). This was followed by many further assessments of the properties of various estimators and test statistics, which continue to be formulated to date. Spatial error autocorrelation is also the only spatial aspect of inference that has been recognized in the standard econometric literature, though only very recently and to a limited extent, e.g., in Johnston (1984) and King (1981, 1987).

In this chapter, I focus in more detail on some salient features of statistical inference when spatial dependence is present in the error term. Since this issue has received considerable attention in the literature, I will avoid an extensive discussion of familiar results. Instead, I will concentrate on a more in depth treatment of a number of aspects that are not always taken into account in a satisfactory way. Specifically, I will focus on tests in nontraditional situations, on the rigorous treatment of the properties of estimators, and on the evaluation of some alternative procedures that have yet to be applied to spatial models.

The chapter consists of three sections. In the first and second, I review some issues related respectively to testing and estimation. In the third section, I consider the robustness of tests for spatial effects when several other sources of misspecification may be present.

8.1. Tests for Spatial Dependence in Regression Error Terms

When the disturbance terms in a regression model show spatial dependence, the standard assumption of a spherical error covariance matrix fails to hold. Instead, this matrix is of a general form $\Omega(\theta)$, typically parameterized in function of a small number of coefficients. In spatial econometrics, these coefficients are associated with a pattern of spatial interaction or spatial structure that is assumed to cause the dependence.

The most commonly used assumption for the form of spatial dependence is a spatial autoregressive specification. For the simple linear regression model

$$y = X\beta + \epsilon$$

this gives, for the error vector ϵ:

$$\epsilon = \lambda W\epsilon + \mu \tag{8.1}$$

where λ is a spatial autoregressive coefficient, W is the usual spatial weight matrix, and μ is an error term that satisfies the classical assumptions of independent identical distribution (i.i.d), with constant variance σ^2. [1] The corresponding error variance is of the form:

$$\Omega = E\ [\epsilon.\epsilon'] = E\ \{[(I - \lambda W)^{-1}\mu][(I - \lambda W)^{-1}\mu]'\}$$

or

$$\sigma^2.\Omega(\lambda) = \sigma^2.(I - \lambda W)^{-1}.[(I - \lambda W)^{-1}]'.$$

A number of alternative specifications have been suggested as well, such as a moving average in the errors in Burridge (1980). Other, less familiar structures were advanced by Fisher (1971), Pocock, Cook and Shaper (1982), and Cook and Pocock (1983). Fisher suggests the use of $\Omega = A^{1/2}W.A^{1/2}$, where A is a diagonal matrix of error variances, and W is a symmetric spatial weight matrix. In the work by Pocock and Cook the spatial dependence in the error term is parameterized as a correlation in function of the distance separating the observations. [2]

In the standard econometric literature, the problem of *spatial* residual correlation is sometimes considered as an issue in the design of survey data, e.g., in Scott and Holt (1982), and King and Evans (1985, 1986). Typically, the form of dependence is without reference to spatial structure and does not acknowledge any distance decay effect. It is formally taken into account in the form of equicorrelated error terms. The resulting block effects are analyzed by means of standard time series approaches, such as a Durbin Watson test.

In the remainder of the chapter, I will consider the nonspherical error variance matrix associated with the spatial autoregressive form (8.1) only. Since the various tests and estimators can be extended to apply to the other forms of residual spatial autocorrelation in a fairly straightforward way, this does not entail a loss of generality.

This section consists of four parts. In the first two, I present a review of tests for residual autocorrelation in the standard regression model, i.e., without a spatially lagged dependent variable or heteroskedasticity. These two special cases are considered in turn in the third and fourth part.

8.1.1. Tests for Residual Spatial Autocorrelation Based on the Moran I Statistic

A Moran I statistic for spatial autocorrelation can be applied to regression residuals in a straightforward way. Formally, this I statistic is:

$$I = [N/S].\{[e'We]/e'e\} \tag{8.2}$$

where e is a vector of OLS residuals, W is a spatial weight matrix, N is the number of observations, and S is a standardization factor, equal to the sum of all elements in the weight matrix. For a weight matrix that is normalized such that the row elements sum to one, expression (8.2) simplifies to:

$$I = e'We/e'e \qquad (8.3)$$

The interpretation of this test is not always straightforward, even though it is by far the most widely used approach. Indeed, while the null hypothesis is obviously the absence of spatial dependence, a precise expression for the alternative hypothesis does not exist. Intuitively, the spatial weight matrix is taken to represent the pattern of potential spatial interaction that causes dependence, but the nature of the underlying stochastic process is not specified. Usually it is assumed to be of a spatial autoregressive form. However, the coefficient (8.3) is mathematically equivalent to an OLS regression of We on e, rather than for e on We which would correspond to an autoregressive process. Moreover, as Burridge (1980) has shown, the Moran test is proportionate to a Lagrange Multiplier test against either a spatial autoregressive or a spatial moving average model.

The asymptotic distribution for the Moran statistic with regression residuals was developed by Cliff and Ord (1972, 1973, 1981). [3] This distribution, for a properly transformed variate, is shown to correspond to the standard normal. The transformation is the usual one:

$$z_I = \{I - E[I]\}/\{V[I]^{1/2}\}$$

where $E[I]$ is the mean, and $V[I]$ the variance of the Moran statistic, derived under the null hypothesis of no spatial dependence. The inclusion of residuals in (8.3) complicates the expressions for the moments. [4] If an underlying normal distribution is assumed for the error term, the following expressions result for the case of a general (non-standardized) weight matrix:

$$E\ [I] = (N/S).tr(MW)\ /(N-K)$$

$$V\ [I] = (N/S)^2.\{tr(MWMW') + tr(MW)^2 + [tr\ (MW)]^2\}/(N-K)(N-K+2)$$

$$- \{E[I]\}^2$$

where W is the weight matrix, N is the number of observations, S is the sum of all elements in the weight matrix (=N for a standardized W), and M is the projection matrix $I - X(X'X)^{-1}X'$.

An alternative, nonparametric approach would yield pseudo significance levels derived from an empirical frequency distribution for the statistic, computed for a large number of permutations of the residuals. More precisely, if N different permutations are considered, a pseudo significance level of $(K+1)/(N+1)$ can be assigned to the result, where K is the number of values in the empirical frequency distribution that are more extreme than the observed statistic. This procedure has attractive finite sample properties, especially when the assumption of normality is not satisfied. Nevertheless, in most applied studies where tests for residual spatial autocorrelation are carried out, the normal approach is typically taken. Under this assumption of normal errors, King (1981) has shown that the Moran test is Locally Best Invariant (LBI) in the neighborhood of $\lambda=0$.

The analysis of spatial dependence in regression residuals is complicated by the fact that these are imperfect estimates for the unobserved error terms. The main difficulty follows from the correlation of the OLS residuals. In order to compensate for this, several other residual estimates have been suggested, which have a scalar covariance matrix. Well known examples are the BLUS (Best Linear Unbiased Scalar), RELUS (Recursive Linear Unbiased Scalar), and LUF (Linear Unbiased with Fixed covariance) residuals.

The effect on the properties of the Moran test of a different choice of estimator for the residuals was investigated by Bartels and Hordijk (1977), and Brandsma and Ketellapper (1979b). In a number of Monte Carlo simulation experiments, using the spatial configuration of 39 regions in the Netherlands and 26 Irish counties, they found that in general the Moran test based on the OLS residuals achieved the highest power.

8.1.2. Tests for Residual Spatial Autocorrelation Based on Maximum Likelihood Estimation

As outlined in detail for the general spatial process model in Section 6.3, the Wald (W), Likelihood Ratio (LR), and Lagrange Multiplier (LM) tests are asymptotic approaches based on maximum likelihood estimation. Tests for the presence of residual spatial autocorrelation can be formulated within this framework as special cases of the general model.

In contrast to the Moran test approach in the previous section, the ML based tests are rigorously structured in terms of specific null and alternative hypotheses. Indeed, the three tests can be considered as different ways of dealing with an omitted variable problem. Formally, the null and alternative hypotheses are: [5]

$$H_0 : \lambda = 0$$

$$H_1 : \lambda \neq 0.$$

The regression model with a spatially autoregressive error term is a special case of the general spatial process model, with parameters $\rho=0$ and $\alpha=0$. Formally, the likelihood for this model is, in the notation of Chapter 6:

$$L = -(N/2).\ln(\pi) -(N/2).\ln(\sigma^2) + \ln |B|$$
$$-(1/2).\sigma^{-2}.(y-X\beta)'B'B(y-X\beta) \tag{8.4}$$

Under the null hypothesis, i.e., with $\lambda=0$, and thus also $B=I$, this becomes the usual likelihood in the linear regression model:

$$L_0 = -(N/2).\ln(\pi) -(N/2).\ln(\sigma^2)-(1/2).\sigma^{-2}.(y-X\beta)'(y-X\beta) \tag{8.5}$$

The likelihood ratio test for residual spatial autocorrelation is based on the difference between (8.4) and (8.5). When the coefficients in the log−likelihoods are replaced by their ML estimates, a much simpler concentrated likelihood results. Specifically, since the estimate for the error variance in either model is 1/N times a properly weighted sum of squared residuals (i.e., the last term in each expression), the last term in (8.4) and (8.5) simplifies to a constant $(-N/2)$. A straightforward difference of the remaining terms yields the LR test as:

$$LR = N.[\ln(\sigma_0^2) - \ln(\sigma_1^2)] + 2\ln.|I - \lambda W| \sim \chi^2(1) \tag{8.6}$$

where σ_0^2 is the estimated residual variance for the model under the null (without residual spatial autocorrelation, i.e., a simple regression) and σ_1^2 is the estimated residual variance for the spatial model. The presence of the Jacobian determinant in (8.6) differentiates this result from the serial correlation case for time series data. [6]

In order to derive the Wald test, the asymptotic variance matrix of the ML estimates in the full model is needed. For the special case considered here, the expression presented in Section 6.2.5. simplifies greatly. Indeed, the resulting information matrix is block diagonal in the regression coefficients and in the error—related parameters σ^2 and λ. [7] The error—related part consists of the following 2 by 2 matrix:

$$\begin{bmatrix} N/2\sigma^4 & \sigma^{-2}.\mathrm{tr}\,W.B^{-1} \\ \sigma^{-2}.\mathrm{tr}\,W.B^{-1} & \mathrm{tr}\,(WB^{-1})^2 + \mathrm{tr}\,(WB^{-1})'(WB^{-1}) \end{bmatrix}$$

A partitioned inversion of this matrix, for the element corresponding to λ, yields:

$$\mathrm{Var}\,(\lambda) = [t_2 + t_3 - (1/N).(t_1)^2]^{-1}$$

with the following simplifying notation:

$$t_1 = \mathrm{tr}\,W.B^{-1}$$

$$t_2 = \mathrm{tr}\,(WB^{-1})^2$$

$$t_3 = \mathrm{tr}\,(WB^{-1})'(WB^{-1})$$

Consequently, a Wald test for residual spatial autocorrelation follows as:

$$W = \lambda^2 . [t_2 + t_3 - (1/N)(t_1)^2] \sim \chi^2(1) \tag{8.7}$$

where the coefficient λ is replaced by its ML estimate. Alternatively, the square root of this expression can be considered as a standard normal variate.

As pointed out before, the Lagrange Multiplier test is based on estimation under the null hypothesis only. This results in an easily implemented statistic, derived from OLS residuals and some additional calculations of weight matrix traces. As in equation (6.36) of Section 6.3.4, this statistic is of the form:

$$LM = (1/T).[e'We/\sigma^2]^2 \sim \chi^2(1) \tag{8.8}$$

where $T = \mathrm{tr}\,\{(W + W').W\}$.

The three ML based tests are asymptotically equivalent, though typically different in finite samples. Very little is known about their small sample behavior.

Some limited evidence was presented by Brandsma and Ketellapper (1979b). They found the LR test to perform poorly compared to the Moran based tests in a Monte Carlo experiment for irregular spatial weight matrices, for regions in the Netherlands. However, a much wider range of experiments would be necessary before general conclusions can be drawn about the power of these tests in situations of interest to the empirical regional scientist.

8.1.3. Testing for Residual Spatial Autocorrelation in the Presence of Spatially Lagged Dependent Variables

In this and the next subsection, two situations are examined more closely, in which a one−directional test against spatial residual autocorrelation (H_0: $\lambda=0$) is carried out in the presence of other known sources of misspecification. The tests are based on the Lagrange Multiplier principle.

The treatment differs from the one−directional tests discussed in Section 6.3.4. There, the other parameters in the spatial process model are a priori set to zero (i.e., $\rho=0$, $\alpha=0$). Here, some of these parameters are included in the model estimation, i.e., they are assumed to be non−zero. The case considered first consists of testing for spatial residual autocorrelation in the presence of a spatially lagged dependent variable. In terms of the coefficients of the general model, this means that $\rho \neq 0$, and $\alpha = 0$.

In the notation of Chapter 6, the model under consideration can be expressed as:

$$B.(Ay - X\beta) = \mu \qquad (8.9)$$

with

$$E [\mu\mu'] = \Omega = \sigma^2.I$$

where A and B are associated respectively with the spatially lagged dependent variable (parameter ρ) and the spatial dependence in the error term (parameter λ). Also, the weight matrices W_1 and W_2 included in A and B do not have to be identical.

Following the usual approach in LM tests, the coefficient vector is partitioned in terms of the parameters considered in the null hypothesis and the other parameters. For model (8.9), this yields:

$$\theta = [\lambda \mid \rho \; \beta' \; \sigma^2]$$

Consequently, under the null hypothesis, with $\lambda=0$, the specification becomes:

$$Ay - X\beta = \mu$$

i.e., a mixed regressive spatially autoregressive model with standard disturbance terms.

In order to implement the LM test, maximum likelihood estimation of model (8.9) is necessary. Although this is carried out by nonlinear optimization, a concentrated likelihood can be used, which reduces to a search over one parameter only (ρ). The resulting optimization problem simplifies greatly and can be carried out in a straightforward way, as illustrated in Section 12.1.

The LM test is based on the score vector and the partitioned information matrix under the null hypothesis. The relevant expressions can be found as special cases of the general formulation in Section 6.2.5, by imposing the constraints $\lambda = 0$, $B = I$, and $\Omega = \sigma^2 I$. As derived in Anselin (1988a), they are:

$$\partial L/\partial\lambda = \sigma^{-2}.(Ay - X\beta)'W_2.(Ay - X\beta) \tag{8.10}$$

and

$$I_{\lambda\lambda} = [T_{22} - (T_{21A})^2.var(\rho)]^{-1}$$

with the following simpifying notation:

$$T_{22} = tr \{W_2.W_2 + W_2'W_2\}$$

$$T_{21A} = tr \{W_2.W_1.A^{-1} + W_2'W_1.A^{-1}\}$$

and var(ρ) as the estimated variance for ρ in the model under the null.

Since $Ay - X\beta$ is the ML residual for model (8.9), expression (8.10) is similar to N times a Moran coefficient in the appropriate residuals.

Using the same general approach as before, the LM statistic for $H_0: \lambda = 0$, is found as:

$$(e'W_2.e/\sigma^2)^2.\{T_{22} - (T_{21A})^2.var(\rho)\}^{-1} \sim \chi^2(1) \tag{8.11}$$

or, equivalently, as:

$$(e'W_2.e/\sigma^2).\{T_{22} - (T_{21A})^2.var(\rho)\}^{-1/2} \sim N(0,1).$$

This statistic can be computed fairly easily from the output of the ML estimation for a mixed regressive spatially autoregressive model, with the additional evaluation of one cross−product and the relevant traces. The LM approach avoids the need for a complex estimation of the model under the alternative hypothesis and thus provides an easy and operational technique to deal with a methodological issue which had eluded a satisfactory solution for a long time. [8] The test is illustrated empirically in Section 12.2.3.

8.1.4. Testing for Residual Spatial Autocorrelation in the Presence of Heteroskedasticity

The second special situation considered here concerns testing for spatial residual autocorrelation in the linear regression model with a pre−specified form of heteroskedasticity. This is relevant when the model takes into account a particular form of spatial heterogeneity, but residual spatial dependence may be suspected as well. For example, the spatial heterogeneity could be a result of spatially varying coefficients or random coefficients, which are discussed in more detail in the next chapter. As before, this situation is a special case of the general approach outlined in Section 6.3.4, but now with $\rho = 0$ and $\alpha \neq 0$.

In the notation of Chapter 6, the model under consideration can be expressed as:

$$\Omega^{-1/2}.B.(y - X\beta) = \nu \tag{8.12}$$

with

$$E\ [\nu\nu'] = I,$$

and Ω embodying a form of heteroskedasticity that can be expressed in function of pre−specified variables Z.

The partitioning of the parameter vector needed for the implementation of an LM test on H_0: $\lambda = 0$ is:

$$\theta = [\lambda \mid \beta' \ \alpha']'$$

Consequently, under the null hypothesis, model (8.12) reduces to the familiar case of a linear regression with heteroskedastic errors:

$$\Omega^{-1/2}.(y - X\beta) = \nu$$

Maximum likelihood estimates for the parameters in this model can be obtained by means of several iterative techniques, or using an explicit nonlinear optimization approach. [9]

As before, the relevant score vector and partitioned information matrix can be found by imposing the constraints $\lambda = 0$, $B = I$ (and, also, $\rho = 0$ and $A = I$) on the expressions given in Section 6.3.4. This yields: [10]

$$\partial L/\partial \lambda = (y - X\beta)'\Omega^{-1}.W_2.(y - X\beta) \tag{8.13}$$

and

$$I\lambda\lambda = \text{tr}\ \{W_2.W_2 + \Omega.W_2'.\Omega^{-1}.W_2\} = T$$

Again, the score vector can be seen as similar to a Moran expression in the ML residuals, i.e., as a cross product between e and W_2.e in the metric Ω^{-1} (i.e., weighted by the inverse diagonal elements of Ω).

In the usual fashion, the Lagrange Multiplier test for H_0: $\lambda = 0$, follows as:

$$[e'\Omega^{-1}.W_2.e]^2\ /\ T \sim \chi^2\ (1) \tag{8.14}$$

or, equivalently:

$$[e'\Omega^{-1}.W_2.e].T^{-1/2} \sim N(0,1)$$

with e as the residuals, and T as above.

This test statistic can be evaluated from the output of the ML estimation of (8.12), in a straightforward way, with the additional computation of one cross product and a trace. It is illustrated empirically in Section 12.2.7.

The LM statistics outlined in this and the previous section provide an attractive way to test for residual spatial autocorrelation in more complex models, since they can be computed from the simpler specification. Although this still necessitates nonlinear optimization, the resulting problem is much more tractable than a solution to the general first order conditions derived in Section 6.2. Also, the tests have clear asymptotic properties. In contrast, this is not the case for the often used ad hoc procedures such as the application of a Moran test under a randomization assumption. However, as with all asymptotic procedures, the performance of the LM tests in finite sample situations is not necessarily satisfactory, particularly when the assumptions underlying the likelihood may not be appropriate. This issue remains to be investigated.

8.2. Estimation in the Presence of Spatially Dependent Error Terms

The linear regression model with spatially autoregressive errors is by far the most relevant spatial specification for applied empirical work on cross−sectional data. Indeed, models with spatially lagged dependent variables tend to have a much narrower scope and are only applied in investigations of specific spatial processes. On the other hand, the spatial dependence in the error term is likely to be present in most data sets collected for contiguous and aggregate spatial units.

In this section I briefly review a number of issues that affect the estimation of the parameters in the regression model, and elaborate further on the general results that were presented in Chapter 6. Specifically, I review the relative merits of OLS, GLS, EGLS and ML estimation procedures. I also evaluate two iterative techniques that were introduced in analogy to the treatment of serial correlation in time series data: the spatial Cochran−Orcutt approach and the spatial Durbin approach.

8.2.1. OLS, GLS, EGLS and ML

As shown above, the spatial dependence in the error term of a linear regression results in a nonspherical error covariance matrix of the form $\sigma^2\Omega(\lambda)$, with

$$\Omega(\lambda) = (I - \lambda W)^{-1}[(I- \lambda W)^{-1}]'$$
$$= [(I - \lambda W)'(I - \lambda W)]^{-1}. \tag{8.15}$$

Consequently, the variance associated with OLS estimates for the coefficients of the model will not have the usual form of $\sigma^2(X'X)^{-1}$, but instead will be a complex function of the parameter λ. Therefore, even though the OLS estimate retains its unbiasedness, inference based on the usual variance estimates may be misleading. Formally, the bias is:

$$E\ [b - \beta] = E\ \{(X'X)^{-1}X'\epsilon\} = 0$$

and the associated variance,

$$E\ [b - \beta][b - \beta]' = E\ \{(X'X)^{-1}X'\epsilon\epsilon'X(X'X)^{-1}\}$$

$$= \sigma^2.(X'X)^{-1}X'[(I-\lambda W)'(I-\lambda W)]^{-1}X(X'X)^{-1} \tag{8.16}$$

If the error covariance is known, the Best Linear Unbiased Estimate is the Aitken Generalized Least Squares. In terms of model (8.1), this implies that both the structure of the spatial dependence (the W) as well as the associated coefficient (λ) should be known. In practice this is typically not satisfied, except in artificial Monte Carlo simulations.

The Estimated GLS consists of an application of the GLS principle with consistent estimates for the parameters in Ω substituted for the unknown population values. In the spatial model, the resulting estimate is:

$$b_{EGLS} = [X'(I-\lambda W)'(I-\lambda W)X]^{-1}X'(I-\lambda W)'(I-\lambda W)y \tag{8.17}$$

which is numerically equivalent to OLS estimation on suitably transformed variables,

$$b_{EGLS} = [X^{*'}X^{*}]^{-1}X^{*'}y^{*}$$

with

$$X^{*} = (I - \lambda W)X$$

$$y^{*} = (I - \lambda W)y$$

and λ replaced by a consistent estimate.

The associated coefficient variance matrix is of the form: [11]

$$Var\ (b_{EGLS}) = \sigma^2.[X'(I-\lambda W)'(I-\lambda W)X]^{-1} \tag{8.18}$$

and

$$\sigma^2 = (y - Xb_{EGLS})'(y - Xb_{EGLS})/N. \tag{8.19}$$

The asymptotic nature of the results for EGLS estimation is a point which is often ignored in applied work. In finite samples, the distribution of the estimates is not well defined, nor is EGLS necessarily superior to OLS in a mean squared error sense. In addition, the properties of the EGLS estimate are sensitive to a correct specification of $\Omega(\lambda)$. In the spatial model, this is primarily determined by the choice of the weight matrix.

In many ways the ML estimate is similar to the EGLS approach, with the difference that the estimation of the error parameter λ is clearly specified and included in the overall inference in a consistent manner. This has some implications for the small sample properties of the ML estimator.

As shown rigorously in Andrews (1986), various estimators for the parameters of the linear regression model with a general error covariance matrix $\Omega(\theta)$ are unbiased in finite samples (in addition to the usual consistency, asymptotic normality, and asymptotic efficiency). The conditions for this result that are particularly relevant in the spatial model pertain to the way in which the estimate for the nuisance parameter is a function of the errors. Specifically, this should be an even function, in the sense that the same estimate for the θ should follow if all error terms have reversed signs. As will be illustrated below, the estimate for λ in the spatial model is essentially found from a minimization of a quadratic form in the errors, which clearly satisfies the conditions for the Andrews result. [12]

It follows from the first order conditions for the maximization of the log likelihood (8.4), as a special case of the derivations in Section 6.2.4, that the EGLS results for β (8.18) and σ^2 (8.19), are also maximum likelihood. [13] After substituting these in (8.4), a concentrated likelihood is obtained which only contains the parameters λ and the residuals $e=(y - Xb_{EGLS})$:

$$L_C \propto -(N/2).\ln\ e'(I-\lambda W)'(I-\lambda W)e +\ln\ |I-\lambda W| \qquad (8.20)$$

Conditional upon the values for b_{EGLS} that gave rise to the residuals, expression (8.20) can be maximized for λ by a simple search procedure. The resulting estimate can be used to derive a new set of b_{EGLS}, which in turn yields new residuals. This process can be iterated until eventual convergence to a maximum for the overall likelihood, as illustrated in Section 12.1.

Two alternative iterative procedures are discussed next.

8.2.2. Iterative Procedures

In analogy to the well known Cochrane–Orcutt (1949) and Durbin (1960) procedures developed for the case with serial error correlation in time series, similar approaches have been suggested for the spatial model, e.g., by Hordijk (1974), Bartels (1979), and Anselin (1980, 1981). [14]

These procedures are intended to provide computational short cuts and to avoid the multiple iterations in the nonlinear maximum likelihood approach.

The spatial Cochrane–Orcutt analogy consists of three steps. In the first, OLS is carried out on the linear model. The residuals are then used to obtain an estimate for the spatial autoregressive coefficient λ, which is subsequently substituted in an EGLS procedure. This can possibly be repeated until convergence.

The initial implementation of this approach was carried out by Hordijk (1974) on the basis of OLS estimation for the spatial parameter. As argued before, the resulting estimate will not be consistent. [15] When the estimate for λ is based on a maximum likelihood algorithm, it becomes numerically equivalent to the first step in an ML estimation for the complete model. Consequently, no gain is achieved in terms of computational simplicity.

The spatial Durbin approach is based on the formal equivalence of the model to a mixed regressive spatial autoregressive specification, in analogy to the suggestion by Durbin for the time series case. Formally, as in the treatment of the general model in Chapter 6:

$$y = X\beta + (I - \lambda W)^{-1}\mu$$

or

$$y = \lambda Wy + X\beta - \lambda WX\beta + \mu \qquad (8.21)$$

where μ is an error term with spherical covariance matrix.

A maximum likelihood estimation or instrumental variables approach to (8.21) will yield consistent estimates for λ and β, either in an unconstrained or constrained procedure. The estimation procedure can be stopped at this point, or a further EGLS can be performed, based on the estimate for λ, as illustrated in Section 12.2.5.

The simplification in (8.21) is further analyzed in the treatment of the common factor problem in Section 13.3. It should be noted that, in contrast to the time series case, the estimation of model (8.21) is actually computationally easier than for the regression with spatially dependent errors in (8.18)−(8.20). [16]

The approaches based on ML or IV estimation have the usual asymptotic properties. In finite samples, no exact results are available. Indeed, the ad hoc procedures and OLS may perform acceptably and even be superior in terms of bias and mean squared error, as illustrated by some limited Monte Carlo experiments in Anselin (1981). However, these results are far from general, and more extensive investigations into the finite sample performance of the various approaches are needed. The main advantage of the maximum likelihood approach is that it allows the application of Wald, Likelihood Ratio and Lagrange Multiplier tests, for which the finite sample properties can be approximated using some recent results of Rothenberg (1984b). The extension of these results to the specific spatial model merits further exploration.

8.3. Robustness Issues

In this section, I briefly consider some aspects of robustness related to inference in models with spatially dependent error terms. These aspects have essentially been ignored in spatial econometrics. Given their relevance for applied work, I will review some recent developments from the (standard) econometric literature and evaluate the extent to which they may be introduced in the spatial case.

First, I deal with the situation where other types of misspecification may be present in the spatial model. Specifically, I outline how a recently developed class of heteroskedasticity−robust tests may be applied to the situation of residual spatial autocorrelation. I also consider some suggestions for specification robust estimates for the error covariance matrix.

8.3.1. Testing for Spatial Dependence in the Presence of Other Misspecifications of Unknown Form

The presence of other forms of misspecification in addition to residual spatial autocorrelation will affect the properties of the various tests and estimators. Particularly relevant in this respect are the potential for non–normality, heteroskedasticity and functional misspecification. When confronted with this issue in applied research, an analyst needs to make a judgement about which of the effects may be important in each particular situation. This will involve a trade–off between making assumptions (few tests) and covering all possibilities (many tests). Also, testing for too few types of misspecification may result in a lack of robustness (and a misleading sense of security), while testing for too many will decrease the power of the respective tests.

This issue has been largely ignored in spatial econometrics, even though it received considerable attention in the standard econometric literature. For example, in dealing with serial correlation in a time series context, the robustness of the traditional Durbin Watson and other tests to the presence of heteroskedasticity was assessed by Harrison and Mc Cabe (1975), and Epps and Epps (1977). The joint effects of functional misspecification and serial correlation are discussed by Ghali (1977), Thursby (1981, 1982), and Godfrey (1987).

One approach to dealing with this issue is to develop tests for the joint presence of several forms of misspecification. Examples in the econometric literature are joint tests based on Box Cox transformation: e.g., for functional misspecification and residual autocorrelation in Savin and White (1978), and functional misspecification and heteroskedasticity in Lahiri and Egy (1981). A general approach based on the Lagrange Multiplier principle was suggested in Jarcque and Bera (1980) and Bera and Jarcque (1982). In spatial econometrics, this approach is exemplified in the LM tests by Anselin (1988a), which were outlined in the previous section.

An alternative strategy consists of developing tests that are robust to the presence of other forms of misspecification. Of particular interest to the issue of spatial dependence and spatial heterogeneity is the general class of heteroskedasticity–robust tests recently suggested by Davidson and MacKinnon (1985a). Its application to the spatial model is discussed next.

8.3.2. A Heteroskedasticity–Robust Test for Residual Spatial Dependence

In applied situations, heteroskedasticity is often likely to be present, although its precise form may be unknown. Consequently, a functional specification of this effect, as required for the LM test of Section 8.1.4. would be speculative at best. Moreover, a wrong choice for this specification will affect the power of the LM test.

In light of this, it would be useful to be able to test for residual spatial autocorrelation in the presence of heteroskedasticity, without having to specify its precise form. Unfortunately, a heteroskedastic error variance matrix which is not expressed in terms of a finite number of parameters cannot be estimated consistently, even under the null hypothesis of no spatial dependence. Indeed, the matrix Ω consists of N parameters, one for each error variance. This leads to an incidental parameter problem, since the number of parameters will increase with the number of observations.

A major breakthrough in this respect was achieved by the result of White (1980), who showed that while Ω could not be estimated consistently, the expression $(X'\Omega X)$ and similar cross products in instrumental variables could. The class of heteroskedastic−robust tests on regression directions developed by Davidson and MacKinnon (1985a) is an extension of this result. Their general framework consists of testing a null hypothesis $\gamma = 0$ in the following setup:

$$H_0 : y = X\beta + \mu$$

$$H_1 : y = X\beta + Z\gamma + \mu$$

where Z is a N by R matrix, γ is a R by 1 column of parameters, and μ is an independent but heteroskedastic error term with $E[(\mu_i)^2] = \sigma_i^2$, bounded for all i.

The test statistic for this case is:

$$DM_1 = y'MZ(Z'M\Omega(u)MZ)^{-1}Z'My \sim \chi^2 (R)$$

where the projection matrix $M = I - X(X'X)^{-1}X'$, and $\Omega(u)$ is a diagonal matrix with the squared OLS residuals (under the restricted model, i.e., under H_0). [17] It turns out that this statistic can be found as N minus the residual sum of squares in an auxiliary regression of

$$\iota = U.M.Z\gamma + errors$$

where ι is a vector of ones, and U is a diagonal matrix of OLS residuals. The test is also extended to instrumental variables estimation and nonlinear situations.

A test for residual spatial autocorrelation can be expressed in this framework using the spatial Durbin approach, as in (8.21):

$$H_0 : y = X\beta + \mu \tag{8.22}$$

$$H_1 : y = X\beta + \lambda Wy - \lambda WX\beta + \mu \tag{8.23}$$

Due to the presence of the lagged dependent variable in (8.23), an instrumental variables estimation needs to be considered. With a matrix of instruments Q, and Z as the N by (K+1) matrix [Wy WX], the IV form for the test could be applied, as N minus the sum of squared residuals in an auxiliary regression of:

$$\iota = UMPZ\gamma + errors$$

where ι is as before, U is a diagonal matrix of OLS residuals, and,

$$M = I - PX(X'PX)^{-1}X'P$$

$$P = Q(Q'Q)^{-1}Q'.$$

However, there are some problems. It is not clear whether K+1 or 1 should be used as the proper degrees of freedom, since the constraint on λ implies the constraints on $\lambda\beta$. Furthermore, with a standardized weight matrix, a separate coefficient for the two constant terms in X and WX cannot be identified, since $W\iota=\iota$. Consequently, perfect multicolinearity will result if this aspect is ignored.

An alternative is to take the constraints on the parameters in the spatial Durbin form explicitly into account in a nonlinear formulation, which can be estimated by nonlinear IV:

$$y = f(\beta,\lambda) + \mu$$

with as relevant partial derivatives,

$$\partial f/\partial\beta = X - \lambda WX$$

$$\partial f/\partial\lambda = Wy - WX\beta.$$

A straightforward extension to the DM results to this case yields the test statistic as: [18]

$$(y - f)'MPF(\lambda)[F(\lambda)'PM\Omega(u)MPF(\lambda)]^{-1}F(\lambda)'PM(y-f)$$

where:

$y - f =$ the OLS residuals in (8.22)

$\Omega(u) =$ a diagonal matrix of squared residuals

$M = I - PF(\beta)[F(\beta)'PF(\beta)]^{-1}F(\beta)'P$, a projection matrix

$P = Q(Q'Q)^{-1}Q'$, with Q as a matrix of instruments,

and $F(\lambda)$ and $F(\beta)$ are the matrices of partial derivatives, evaluated under the null hypothesis of $\lambda =0$. For the spatial model, these partial derivatives simplify to:

$$F(\beta) = X$$

$$F(\lambda) = Wy - WXb$$

with b as the OLS estimate. Consequently, the auxiliary regression is of the form:

$$\iota = UMPF(\lambda)\gamma + error$$

where U is a diagonal matrix of OLS residuals, and $MPF(\lambda)$ are the residuals in a regression of $PF(\lambda)$ on $PF(\beta)$, or, of $P(Wy - WXb)$ on PX. The statistic is N minus the residual sum of squares in the auxiliary regression, and is asymptotically distributed as χ^2 with 1 degree of freedom. The performance of this approach in finite samples remains to be investigated.

8.3.3. Robust Error Covariance Matrix Estimates

An alternative approach to inference in the presence of residual spatial dependence of unknown form can be based on some asymptotic results for the case of heteroskedasticity. However, its application is limited to spatial processes for regular lattice structures with well-defined spatial autocorrelation functions. As will be outlined below, an analogy to some of the robust approaches developed in a time series context is conditional upon the existence of a parameterization of the spatial dependence in terms of a lag length. [19]

The heteroskedastic case is based on the result of White (1980) where an asymptotic estimate for the covariance matrix of the OLS estimator is presented, which is consistent in the presence of heteroskedasticity of unknown form. The standard OLS variance for this situation,

$$(X'X)^{-1}X'\Omega X(X'X)^{-1}$$

is estimated by

$$(X'X)^{-1}X'SX(X'X)^{-1}$$

where S is a diagonal matrix of squared regression residuals. Alternatively, the expression $X'\Omega X$ is estimated by a sum of N K by K matrices, $\Sigma_i \ (x_i'u_i).(u_i x_i)$, where u_i is the residual associated with observation i, and x_i is a row vector of explanatory variables for i. [20]

An extension to serial autocorrelation consists of including a sum of matrices of sample autocovariances for lag lengths up to a given limit, of the form

$$\Sigma_s \ \Sigma_t \ (x_t'u_t)(u_{t-s}x_{t-s}) + (x_{t-s}'u_{t-s})(u_t x_t)$$

where t is the time period and t−s the appropriate lag. In principle, the sum is taken over all cross products for which the residual covariances are nonzero. In practice, the choice of the maximum lag is not arbitrary, but related to the admissible degree of dependence and heterogeneity in the underlying process. [21]

This result is still largely theoretical, and its implementation in realistic data situations not unambiguous. The extension of this approach holds promise for spatial processes with a regular structure and well-specified spatial autocorrelation functions. From a purely formal perspective, the replacement of a lag in time by a lag in space (or time−space) can be carried out in a straightforward way, provided that the underlying process satisfies the various (strict) regularity conditions. For processes defined on irregular spatial configurations, this approach does not seem to hold much promise.

A different perspective is provided by the jackknife estimate of variance, which was applied to the heteroskedastic case by MacKinnon and White (1985).

116

This estimate is similar to the bootstrap discussed in Section 7.3. In essence, the least square estimation is repeated N times, each time on a data set from which one observation is dropped. An estimate for the covariance of the OLS estimate is obtained from

$$[(n-1)/n].\Sigma_i \ [b(i) \ - \ (1/N) \ \Sigma_j \ b(j)][b(i) \ - \ (1/N) \ \Sigma_j \ b(j)]'$$

where b(i) is the estimate on the data set without observation i. [22]

An extension of this approach to a general error covariance matrix which also incorporates spatial dependence seems straightforward. However, the properties of such an approach for realistic spatial data sets remain to be investigated. Although few general results have been obtained to date, the potential benefits of a robust approach to inference in spatial models are considerable, given the non—standard properties of most data encountered in applied work.

NOTES ON CHAPTER 8

[1] This model is a special case of the general specification introduced in Section 6.2, with $\rho=0$ and $\alpha=0$.

[2] This approach to spatial autocorrelation leads more directly to the use of spatial correlograms, in analogy to time series analysis. See also Granger (1969), for a similar viewpoint.

[3] In Cliff and Ord (1972), a finite sample approach is presented as well, based on an analogy with the Durbin Watson bounds in time series analysis. However, in contrast to the time series context, the generality of the bounds is limited. In essence, each different spatial weight matrix would necessitate the computation of a new set of bounds. Therefore, in most applications, the asymptotic approach is taken.

[4] For a detailed and a more rigorous derivation, the reader is referred to the texts by Cliff and Ord (1973, 1981) and Upton and Fingleton (1985). The essential difference between the approach for regression residuals and that for other variates is the existence of dependence in residuals. Indeed, it is well known that the OLS residuals are related to the error terms as $e = M\epsilon$, where M is an idempotent projection matrix $I-X(X'X)^{-1}X'$. Consequently, the e will not be uncorrelated, even when the ϵ are. An alternative approach is to ignore the special nature of residuals and to use a randomization framework. The moments of the Moran statistic for that case are given in Cliff and Ord (1981, p. 21).

[5] For many statistical results, the alternative hypothesis is of the form of a local alternative. This formally expressed as $H_1 : \lambda = 0 + \delta$, for which $\delta \rightarrow 0$ with $N^{-1/2}$.

[6] For a more detailed derivation, see Anselin (1980, pp. 130–2).

[7] This is a special case of the general structure in Magnus (1978), and the invariance result presented in Breusch (1980).

[8] For example, see the comments in Cliff and Ord (1981), and Upton and Fingleton (1985).

[9] Overviews are given in Magnus (1978), and Raj and Ullah (1981). Estimation in these models is further considered in Section 9.4.

[10] For a detailed derivation, see Anselin (1988a).

[11] Since the variance matrix is asymptotic, the estimate for σ^2 in (8.19) can be obtained by dividing the sum of squared residuals by either N or N–K (for very large N, the difference between the two will become negligible). In finite samples, the latter is less likely to lead to overly optimistic (small) estimates of variance.

[12] Specifically, this is assumption A.3. (Andrews 1986, p. 692). The other conditions relate to the estimator for β and are clearly satisfied by the ML approach.

[13] See the detailed derivations in Ord (1975), Hepple (1976), Anselin (1980, 1981), Cliff and Ord (1981), and Upton and Fingleton (1985).

[14] An ad hoc first differencing procedure was suggested by Martin (1974). In it, OLS is carried out on transformed variables $y^* = y - Wy$ and $X^* = X - WX$. For a standardized weight matrix, the implied value of $\lambda = 1$ would result in an unstable spatial process. Also, overall, this adjustment does not have clear properties besides the obvious computational ease.

[15] In his empirical application, Hordijk reports a lack of convergence of this approach, which may be due to the inconsistency of the estimate for λ.

[16] This is due to the fact that in every iteration an adjusted b_{EGLS} results in a new nonlinear optimization for λ. In model (8.21) λ and β are estimated jointly, since the value of λ does not depend on the estimate for β. Therefore, instead of many nonlinear searches, only one needs to be carried out. This is further illustrated in Section 12.1.

[17] Davidson and MacKinnon (1985a) consider various forms of $\Omega(u)$, but the one mentioned here shows superior power in finite samples.

[18] For a detailed derivation, see Anselin (1988b).

[19] This strict requirement may potentially be replaced by a parameterization in function of a distance metric, as in Granger (1969) and Cook and Pocock (1983). In either case, consistency will require a large number of observations in each distance–class.

[20] See White (1980, 1984) for details. The application of this approach to other than the standard regression context is discussed more extensively in, e.g., Cragg (1983), Hsieh (1983), Nicholls and Pagan (1983), Chesher (1984), and Robinson (1987). However, most of those extensions are based on the assumption that OLS provides a consistent estimate for the model parameters. As shown in Section 6.1, this is not the case in the specifications with spatially lagged dependent variables. Adjustments that result in better finite sample performance are discussed in MacKinnon and White (1985), and Chesher and Jewitt (1987).

[21] For a rigorous discussion, see White (1984), and White and Domowitz (1984). An extension and suggestions for operational implementation are given in Newey and West (1987).

[22] For details, see Anselin (1988b).

CHAPTER 9

SPATIAL HETEROGENEITY

Many phenomena studied in regional science lead to structural instability over space, in the form of different response functions or systematically varying parameters. In addition, the measurement errors that result from the use of ad hoc spatial units of observation are likely to be non—homogeneous and can be expected to vary with location, area or other characteristics of the spatial units.

To the extent that these aspects of heterogeneity can be related to spatial structure, or are the result of spatial processes, I have designated them by the term *spatial heterogeneity*. This includes familiar econometric problems such as heteroskedasticity, random coefficient variation and switching regressions.

In this chapter, I will discuss some issues of heterogeneity that are of particular interest in spatial econometrics. Specifically, I will deal with the effect of spatial dependence on standard tests for forms of heterogeneity, and with various types of spatial parameter variation that have been suggested in the literature. Since many features of heterogeneity are easily taken into account by the use of standard econometric techniques, I will concentrate on those aspects which have a special spatial flavor, and mainly refer to the literature for the rest.

The chapter consists of four sections. In the first, I briefly outline some general issues associated with spatial heterogeneity. In the second, I focus more specifically on the effect of spatial autocorrelation on tests for heteroskedasticity and structural stability. The third section outlines and evaluates the spatial expansion method of Casetti, as an example of a method which has been suggested to deal with parameter variation over space. In the fourth section I review a number of other techniques that have been proposed to take into account spatial heterogeneity.

9.1. General Aspects of Spatial Heterogeneity

Spatial heterogeneity has been taken into account in a variety of ways in empirical work in regional science. For example, systematic variation of parameters with location is accounted for in the spatial expansion method of Casetti (1972, 1986), and random parameter variation in spatial data is analyzed in urban density studies by Kau and Lee (1977), Johnson and Kau (1980), and Kau, Lee and Sirmans (1986). Structural change of a discrete nature, expressed in the form of switching regressions has been implemented in the work of Brueckner (1981, 1985, 1986), and Kau, Lee and Chen (1983). Instances where heteroskedasticity is incorporated are the urban analyses of Greene and Barnbock (1978), and Anselin and Can (1986).

In general terms, there are two distinct aspects to spatial heterogeneity. One is structural instability as expressed by changing functional forms or varying parameters. [1] The other aspect is heteroskedasticity, which follows from missing variables or other forms of misspecification that lead to error terms with non—constant variance. Ignoring either aspect has well—known consequences for the statistical validity of the estimated model: biased parameter estimates (but not in the presence of heteroskedasticity only), misleading significance levels, and suboptimal forecasts.

The extent of spatial heterogeneity than can be formally incorporated in a model is limited by the incidental parameter problem, i.e., the situation where the number of parameters increases directly with the number of observations. To avoid this, the heterogeneity needs to be expressed in terms of a few distinct categories or parameters. For models with varying coefficients, this implies that the variation should either be determined systematically, in function of a small number of additional variables (as in the spatial expansion method), or stochastically, in terms of an a priori distribution (as in the random coefficient approach). In the case of structural instability of the functional form, the number of different regimes that can be efficiently estimated is limited by degrees of freedom considerations.

In many situations, the basis for the specification of the particular form of heterogeneity in spatial models can be derived from regional science theory. In particular, theories of regional structure and urban form can provide insight into characteristics of spatial data sets that are likely to cause heterogeneity, as well as provide important variables that determine its form.

A complicating factor in spatial analysis is that the misspecifications and measurement errors that may lead to heteroskedasticity, such as the problems with the choice of a spatial unit of observation, are also likely to cause spatial autocorrelation. It is therefore important to consider the effect of the presence of one type of misspecification on the tests for and estimation of the other. Testing for spatial autocorrelation in the error terms in the presence of heteroskedasticity was discussed in the previous chapter. In this chapter the other combination is considered, i.e., the effect of spatial autocorrelation on tests for heteroskedasticity and structural stability. I turn to this issue next.

9.2. Testing for Heterogeneity in the Presence of Spatial Dependence

The effect of serial error autocorrelation in time series on tests for heterogeneity has received some attention in the standard econometric literature. For example, in Epps and Epps (1977), the validity of the Glejser and Goldfeld and Quandt tests was found to be affected by first order autoregressive error terms. [2] Similarly, in Consigliere (1981) and Corsi, Pollock and Prakken (1982), serial autocorrelation is shown to invalidate the results of the well—known Chow test for structural stability. [3]

The analogue of these issues for spatial dependence has been largely ignored. Therefore, in this section, I will discuss some of these effects in more detail, in the context of some well—known tests for heteroskedasticity and structural stability. I also present some alternative formulations in which spatial error autocorrelation is taken into account.

9.2.1. Testing for Heteroskedasticity in the Presence of Spatial Dependence

When the error terms fail to be independent, the distributional properties of several parametric tests for heteroskedasticity are no longer valid. More precisely, this is due to the use of the characteristics of quadratic forms in independent normal variates as the basis for deriving the asymptotic distribution of most test statistics. Consequently, in the absence of independence these results will no longer hold.

In finite samples, no analytical results are available, and the evaluation of the various tests needs to be based on Monte Carlo experiments. In Anselin (1987b) some simulation results are presented on the effect of spatial autocorrelation in the error terms on the bias and power of the Glejser, Breusch—Pagan and White tests. The spatial dependence is in the form of a first order autoregression for a simple standarized contiguity matrix on regular lattice structures, for sample sizes of 25, 50 and 75. Similar to the results for serial autocorrelation, the performance of the tests is seriously affected.

The simulations illustrate how under the null hypothesis of no heteroskedasticity, the empirical rejection frequencies for both the Glejser and Breusch—Pagan tests exceed the nominal significance level when spatial autocorrelation is present. Especially for large positive spatial autocorrelation, this effect is pronounced, yielding rejection frequencies of two to three times the nominal level. The Breusch—Pagan test in particular is sensitive to this. For the White test, the effect is less important, and seems to work in the opposite direction, i.e., lower rejection frequencies when spatial autocorrelation is present.

The power of the tests is affected as well. It tends to be lowered by the presence of large autocorrelation, particularly for positive values of the autoregressive parameter. However, the relative ranking of the three tests does not seem to be sensitive to spatial autocorrelation. Overall, the Glejser test is shown to be the most powerful, with the White test having very poor results. Although the generality of these results is limited by the scope of the experiments, they give a clear indication that caution is needed when interpreting tests for heteroskedasticity in the presence of spatial autocorrelation in the errors. [4]

As pointed out earlier, the general model presented in Section 6.2. encompasses both types of potential spatial effects. Two testing strategies can be suggested, based on the results of maximum likehood estimation for special cases of this general model. One would consist of a set of sequential tests, first for the joint possibility of heteroskedasticity and spatial autocorrelation, and next for one or both of these effects in isolation. The joint test is a special case of the general Lagrange Multiplier approach discussed in Section 6.3.4. and consists of the sum of a Breusch—Pagan statistic and an LM test against spatial residual correlation:

$$(1/2).f'Z(Z'Z)^{-1}.Z'f + (1/T)[e'We/\sigma^2]^2 \sim \chi^2(P+1)$$

where, as before,

$$f_i = (\sigma^{-1}.e_i)^2 - 1$$

$$T = tr[W'W + W^2]$$

and, e is a vector of OLS residuals e_i; σ^2 is the ML variance based on OLS residuals; Z is an N by (P+1) matrix of a constant term and the variables that cause heteroskedasticity; and W is the usual spatial weight matrix.

A significant rejection of the joint null hypothesis could be followed by a test for each of the special cases. In the process of carrying out this sequential procedure, the critical levels used as the basis for rejection of the respective null hypotheses should be adjusted to obtain a correct assessment of the multiple comparisons. For example, this could be achieved by using Bonferroni bounds, which would consist of dividing the overall desired significance level by the number of comparisons. [5]

An alternative is to derive an explicit test for heteroskedasticity in the presence of spatial autocorrelation. Again, this is most easily achieved in an asymptotic framework by a Lagrange Multiplier approach. [6]

As in Section 8.1.4. the starting point is the following model:

$$\Omega^{-1/2}.B.(y - X\beta) = \nu$$

with

$$E\left[\nu\nu'\right] = I$$

$$B = (I - \lambda W)$$

and Ω embodying a form of heteroskedasticity expressed in function of pre−specified variables in the matrix Z. In contrast to the formulation in Section 8.1.4. the partitioned parameter vector for the LM test on H_0: $\alpha = 0$ is:

$$[\ \alpha\ |\ \beta\ \sigma^2\ \lambda\ ].$$

Consequently, under the null hypothesis, the model reduces to the case with spatial autocorrelation in homoskedastic errors, and the error covariance is of the form $\sigma^2.(B'B)^{-1}$. The test is based on maximum likelihood estimates for the parameters under the null, which can be obtained in the usual fashion following the expressions in Sections 8.2. and 12.1.

An application of the same principles as for the general case in Section 6.3.4. yields a test statistic of the following form:

$$LM = (1/4\sigma^4).f'Z.I^{-1}.Z'f$$

where f and Z are as before, but expressed in terms of the spatially weighted ML residuals $e = B(y - X\beta)$, and the variance σ^2.

The inverse term pertains to the relevant partitioning of the information matrix, for the elements of α. Since this matrix is not block diagonal in the elements σ^2, λ and α, no simple expression is available. [7] However, under the null hypothesis, the relevant expression for the inverse of the submatrix for σ^2 and λ is found as the estimated variance for those coefficients. Partitioned inversion of the submatrix in α and the two by two submatrix in σ^2 and λ yields the following expression: [8]

$$LM = (1/2).f'Z.[Z'D.Z]^{-1}.Z'f \sim \chi^2(P)$$

where

$$D = I - (1/2\sigma^4).d.V.d'$$

with

$$d = [\ \iota\ \ 2\sigma^2.w\ ]$$

and, ι is a N by 1 vector of ones; w is a vector consisting of the diagonal elements of $W.B^{-1}$; and, V is the estimated covariance for σ^2 and λ.

In contrast to the Breusch—Pagan result, this test statistic does not have a direct interpretation in terms of the R^2 of an auxiliary regression, since the matrix $Z.[Z'D.Z]^{-1}Z'$ is not idempotent.

In general terms, this expression allows some intuitive insight into the effect of spatial autocorrelation on the traditional statistic. Since the matrix D is positive definite, the resulting LM statistic will tend to be smaller than the one which ignores spatial effects. [9] Consequently, the latter will tend to reject the null hypothesis more often than warranted. This is in general agreement with the results of the Monte Carlo simulations mentioned earlier, although the specific small sample performance of this approach remains to be investigated. An empirical illustration is provided in Section 12.2.4.

9.2.2. Testing for Structural Stability in the Presence of Spatial Dependence

A simple model of structural instability is the situation where the regression parameters take on distinct values in subsets of the sample. In applied regional science, this could easily be the case, for example, when data are used for both recently settled and older metropolitan areas, for urban and rural counties, or for central city and suburban census tracts. A well—known test for the presence of this structural change is the Chow test, based on an F statistic in function of restricted and unrestricted sums of squared residuals.

Formally, the null and alternative hypothesis can be expressed as follows:

$$H_0 : y = X\beta + \epsilon$$

$$H_1 : y = \begin{bmatrix} X_i & 0 \\ 0 & X_j \end{bmatrix} \begin{bmatrix} \beta_i \\ \beta_j \end{bmatrix} + \epsilon$$

where the N by $K_{i(j)}$ matrices X_i and X_j, and K by 1 vectors β_i and β_j are subsets of observations on the explanatory variables with associated regression coefficients. Provided that enough observations are available in each subset, the test is based on the residuals for a regression under the null hypothesis, e_R (restricted estimates), and for a regression under the alternative, e_U (unrestricted estimates), as:

$$C = \{(e_R'e_R - e_U'e_U)/K\}/\{e_U'e_U/(N-2K)\} \sim F(K, N-2K)$$

As shown in Consigliere (1981) and Corsi, Pollock and Prakken (1982) for serial autocorrelation, this test becomes invalid when the error terms ϵ are no longer assumed to be independent. The same result also holds for spatial autocorrelation in the error terms. As a consequence, a test for structural stability in the presence of spatial dependence in the error terms cannot be based on the finite sample F−test, but instead needs to be derived from asymptotic procedures such as the Wald, Likelihood Ratio and Lagrange Multiplier statistics. The main difference between these approaches will be computational, since they are asymptotically equivalent.

In essence, this situation is a special case of tests on linear constraints on the parameters in a generalized least squares framework with error covariance $\sigma^2\Psi$. The proper tests can be expressed as:

$$(e_R'\Psi^{-1}e_R - e_U'\Psi^{-1}e_U) / \sigma^2 \sim \chi^2(K)$$

where the e are ML residuals, σ^2 is the ML estimate for the error variance in the unrestricted model, and the Ψ is replaced by a consistent estimate. [10]

When the error terms follow a spatial autoregressive process, the associated variance matrix is of the form: [11]

$$\Psi = [(I-\lambda W)'(I-\lambda W)]^{-1}$$

The corresponding test statistic becomes:

$$\{e_R'(I-\lambda W)'(I-\lambda W)e_R - e_U'(I-\lambda W)'(I-\lambda W)e_U\} / \sigma^2 \sim \chi^2(K)$$

where λ represents the ML estimate for the spatial parameter and σ^2 the estimate for the error variance for either the restricted model (LM test), the unrestricted model (W test), or both (LR test). Strictly speaking, this approach has only asymptotic validity and the interpretation of the alternative forms of the test may lead to conflicts in finite samples, as discussed in Section 6.3.5. The spatially adjusted tests for structural stability are illustrated empirically in Section 12.2.8.

9.3. Spatial Expansion of Parameters

The spatial expansion method is an approach to dealing with heterogeneity in regression analysis which has received considerable attention in empirical research in geography. This technique, originally suggested by Casetti (1972) in the context of spatially varying parameters, has since been extended to a general framework for model development (Casetti, 1986). In addition, the expansion method has recently been applied to a wide range of empirical problems in urban and regional analysis, e.g., in Jones (1983, 1984), Brown and Jones (1985), and Casetti and Jones (1987, 1988).

In this Section, I discuss some methodological aspects that pertain to estimation and specification searches in the context of the expansion method. These more technical issues are typically ignored in empirical implementations, but should be taken into account to ensure correct inference. Specifically, after a brief formal outline of the approach, I focus on the relationship between the heterogeneity implied by the spatially expanded parameters and the related properties of heteroskedasticity and spatial autocorrelation.

9.3.1. General Principle of the Expansion Method

From an econometric viewpoint, the spatial expansion method can be considered as a special case of systematically varying coefficients in a regression model. The heterogeneity in the phenomenon under study is taken to be reflected in parameter values that differ for each observation. In addition, this differentiation is assumed to be expressed as a function of a number of auxiliary variables, which leads to a more complex model formulation. In the terminology of the expansion method, the original simple homogeneous specification is called the initial model, whereas the complex heterogeneous formulation is called the terminal model.

In the early applications of the expansion method, the auxiliary variables consisted of trend surface polynomials in terms of the coordinates of the locations of the observations, hence the characterization as *spatial* expansion. More recently, this has been generalized to orthogonal principal components of trend surface expressions and other more complex formulations.

Without loss of generality, the properties of the spatial expansion method can be more formally illustrated for a simple regression with one explanatory variable. The initial model is then:

$$y = \beta_0 + \beta_1 x + \epsilon$$

where β_0 and β_1 are regression coefficients, and x is a vector of observations on the explanatory variable. The heterogeneity is reflected in a lack of stability of the parameters over the observational units. This is incorporated in the model by assuming each individual parameter (or a subset of the parameters) to be an exact function of a finite number of expansion variables, e.g., z_1 and z_2 in:

$$\beta_1 = \gamma_0 + \gamma_1 . z_1 + \gamma_2 . z_2$$

Substitution of the expanded parameter β_1 into the original formulation yields the terminal model, as:

$$y = \beta_0 + (\gamma_0 + \gamma_1 . z_1 + \gamma_2 . z_2) . x + \epsilon$$

or,

$$y = \beta_0 + \gamma_0 . x + \gamma_1 . (z_1 . x) + \gamma_2 . (z_2 . x) + \epsilon$$

If the expanded terminal model is indeed the correct specification, then the parameter estimates in the initial model will be biased, due to the omitted variable problem. More specifically, and using a well–known result from partitioned regression, the OLS estimates for the coefficients in the initial model can be expressed as:

$$b = (X'MX)^{-1}X'My$$

with

$$M = I - Z(Z'Z)^{-1}Z'$$

and b as the estimate for the coefficient subvector $[\beta_0 \ \gamma_0]$, $X = [\iota \ x]$, and $Z = [z_1 . x \ z_2 . x]$. In terms of expected values, it can be shown, after some straightforward applications of partitioned inversion, that:

$$E[b] = \beta + (X'X)^{-1}X'Z\gamma$$

where γ is the population parameter subvector corresponding to $[\gamma_1 \ \gamma_2]$. The elements of the matrix Z are products of the elements of X with the expansion variables. Therefore, X and Z will not be orthogonal, and thus the OLS estimates for the initial model will be biased.

In a typical empirical application of the expansion method, the degree of the appropriate trend surface polynomial is determined from a series of stepwise regressions. Although often ignored in practice, the problems associated with this type of ad hoc specification search are serious, and tend to invalidate the formal probabilistic framework on which the inference is based. [12] Furthermore, the expanded variables will tend to be highly multicolinear, which will lower the precision of the estimates. [13]

In sum, the spatial expansion method provides a simple and attractive way to take into account heterogeneity in the coefficients of regression models. It also provides an explicit spatial representation of the instability in parameters. However, its implementation needs to be carried out with caution, especially when there are no good a priori reasons to guide the choice of the expansion variables. Some other important implementation issues are discussed next.

9.3.2. Spatial Expansion and Heteroskedasticity

In applied work, the assumption of an exact relationship between the coefficients and their spatial expansions is hard to maintain. Indeed, the very search for the proper degree for a trend surface expansion, and the use of orthogonal principal components imply the existence of a stochastic error term.

In more precise terms, this leads to the following expansion (in the same notation as before):

$$\beta_1 = \gamma_0 + \gamma_1 . z_1 + \gamma_2 . z_2 + \mu$$

where μ is a stochastic error term, which could be assumed to be normally and independently distributed with variance σ_u^2. Substitution of this expression into the initial model yields a different terminal model:

$$y = \beta_0 + \gamma_0.x + \gamma_1.(z_1.x) + \gamma_2.(z_2.x) + \mu.x + \epsilon$$

or, with a new error term

$$\omega = \mu.x + \epsilon$$

$$y = \beta_0 + \gamma_0.x + \gamma_1.(z_1.x) + \gamma_2.(z_2.x) + \omega$$

which is a model with a heteroskedastic error. [14]

With spherical disturbances in the initial model, the error variance in the terminal model is a function of the explanatory variables x, as:

$$var(\omega) = \sigma_u^2.x^2 + \sigma_e^2.$$

provided that the expansion error and the model error are independent, i.e., $E[\mu_i.\epsilon_i]=0$, which can reasonably be assumed in most instances.

Consequently, inference about the parameters in the terminal model which ignores the heteroskedastic character of the error terms may be misleading. This is important in this context, since the significance of the parameters γ_1 and γ_2 determines the evaluation of structural instability. Although an implementation of EGLS would be straightforward, correct inference could still be based on OLS estimates, but the coefficient variance needs to be properly adjusted. For example, this could be carried out according to one of the asymptotic heteroskedastic—consistent procedures discussed in Section 8.3, and illustrated in Section 12.2.6.

The need to resort to an asymptotic approach may result in further complications. As before, there is the additional uncertainty about finite sample approximations. Moreover, the validity of the asymptotic results is based on a number of boundedness regularity conditions which may not be satisfied in the expanded models. In particular, the form of heteroskedasticity in the model should preclude an infinite error variance as the sample approaches infinity. For example, this would exclude the coefficients from a trend surface model to be expanded in terms of other variables, since the ensuing heteroskedastic component, as a function of powers of coordinates, could potentially become infinitely large. [15]

These issues are typically ignored in empirical applications of the spatial expansion method, although they are crucial in ensuring the correct interpretation of the estimation results.

9.3.3. Spatial Expansion and Spatial Autocorrelation

A final implementation issue associated with this approach is the extent to which a spatial expansion of parameters eliminates spatial autocorrelation in the errors of the initial model, as shown in Jones (1983) and Casetti and Jones (1988). In these two instances, the authors found that a Moran test for spatial autocorrelation in the residuals of a model became insignificant after expansion of the coefficients.

In order to gain a better insight into this issue, consider the OLS estimate for β in the initial model, when the expanded terminal model is the true specification. Consequently, the initial model will be misspecified, and the OLS estimate is biased:

$$E[b] - \beta = (X'X)^{-1}X'Z\gamma$$

As a result, the estimated residual is not the usual expression, but, in terms of the population parameters:

$$
\begin{aligned}
y - Xb &= Z\gamma + \epsilon - X(X'X)^{-1}X'Z\gamma - X(X'X)^{-1}X'\epsilon \\
&= M(Z\gamma + \epsilon)
\end{aligned}
$$

A Moran statistic based on the residuals in the misspecified model could potentially indicate significant spatial autocorrelation, even when the error terms are independent. This is similar to the situation in time series analysis, where tests for serial autocorrelation, and the Durbin Watson test in particular, have been shown to exhibit power against a variety of misspecifications. An uncritical interpretation of test results may lead to a false consideration of serial dependence, whereas other forms of misspecification, such as omitted varables or nonlinearities may be the correct underlying cause. [16]

This can be illustrated in general terms for the spatial case by considering the expected value for a Moran I coefficient based on the residuals in the misspecified model:

$$I = u'Wu/u'u$$

where $u = M(Z\gamma + \epsilon)$. Following the same general reasoning as in Cliff and Ord (1981, pp. 201−203), the expected value can be found as the ratio of the expected values of the numerator and denominator. The latter is $\sigma^2.(N - K)$, with K as the number of coefficients in the model. The expected value for the numerator is:

$$
\begin{aligned}
E[u'Wu] &= E\{(Z\gamma + \epsilon)'MWM(Z\gamma + \epsilon)\} \\
&= \gamma'Z'MWMZ\gamma + \sigma^2.\mathrm{tr}(MW) \\
&= \gamma'Z'MWMZ\gamma - \sigma^2.\mathrm{tr}(X'X)^{-1}X'WX.
\end{aligned}
$$

Consequently, $E[I]$ becomes:

$$E[I] = \gamma'Z'MWMZ\gamma/[\sigma^2(N-K)] - \mathrm{tr}(X'X)^{-1}X'WX/(N-K).$$

Since the first term in this formulation is positive, the expected value will exceed the result for a properly specified model, which is the second term in the expression. Intuitively, since too small a value is subtracted from the raw I measure to obtain the standardized z coefficient, the Moran test in the misspecified model may be more likely to reject the null hypothesis of no spatial autocorrelation.

The extent to which this is the case depends directly on the degree of coefficient instability (the $Z\gamma$).

In other words, the relation between spatial expansion and spatial autocorrelation has to be stated carefully. The issue is not whether spatial expansion in the parameters eliminates spatial autocorrelation in the error terms. Rather, it turns out that the Moran test may have power against misspecifications of the form implied by spatial coefficient instability.

9.4. Other Forms of Spatial Heterogeneity

In this section, I will consider four alternative procedures which have been suggested at times to take into account spatial variation. The first two approaches, random coefficient estimation and error components models, were proposed in Arora and Brown (1977) as *solutions* to the problem of spatial autocorrelation. As I will argue in more detail below, this is an overstatement, since spatial dependence is not addressed by these techniques. Rather, these methods are well suited to take into account heterogeneity of a general unspecified form, which could have a particular *spatial* interpretation, although this is not necessary.

The third approach considered in this section deals with variations in the functional specification over the sample space, in the form of switching regressions. Again, whether or not this results in a meaningful spatial interpretation depends on the particular context. For example, in some recent applications of urban density modeling by Brueckner (1985, 1986), the different regimes are related directly to a vintage model of urban structure.

The final method discussed in this section is the spatial adaptive filtering approach developed by Foster and Gorr (1983, 1984, 1986), as an application of adaptive estimation to the spatial domain.

9.4.1. Random Coefficient Variation

In many empirical situations, no obvious variables are available to determine a specific form for the spatial variation in the regression coefficients. In such cases, an alternative approach is the Hildreth–Houck (1968) random coefficient model, where the coefficients for individual observations are considered as random drawings from a multivariate distribution. Formally, for each observation i: [17]

$$y_i = x_i'\beta_i$$

where x_i' is a row vector of explanatory variables, and β_i is a K by 1 vector of coefficients, determined as:

$$\beta_i = \beta + \mu_i$$

with the error μ_i typically assumed to be distributed according to a K−variate normal distribution. This distribution has a mean 0 and a covariance matrix Σ, which can be diagonal in the special case of no covariance between the errors of the individual parameters. A straightforward substitution yields a regression relationship of the usual form, with a heteroskedastic error term ν_i:

$$y_i = x_i'\beta + \nu_i$$

where

$$\nu_i = x_i'\mu_i$$

and

$$E[\nu_i] = 0$$

$$V[\nu_i] = x_i'\Sigma x_i.$$

The model parameters can be estimated using a maximum likelihood approach, or by a variety of iterative EGLS procedures. In applied situations with a poorly specified model, a problem may occur when the estimates for the elements of Σ do not yield a positive definite covariance matrix. [18] The potential for random coefficient variation can be assessed before estimating the more complex specification, by means of tests for this particular form of heteroskedasticity, such as the Breusch–Pagan (1979) test.

Clearly, in this form the random coefficient model is a special case of heteroskedasticity, and therefore it does not address the issue of spatial dependence. [19]

In order to incorporate spatial autocorrelation, a more complex specification is needed. For example, this could be achieved by separating the randomness of the intercept from that of the other coefficients, and thus excluding interaction between the error for the constant term and the other random elements. The intercept error could then be allowed to reflect spatial dependence of the usual form, for example, as an autoregressive process.

Formally, the model can be expressed as:

$$y_i = \alpha + x_i'\beta_i + \epsilon_i$$

where the $x_i'\beta_i$ is as before and ϵ_i is the disturbance associated with the intercept. The overall model error can be expressed in vector form as:

$$\omega = \epsilon + \nu$$

with ν as before, and

$$\epsilon = \lambda W \epsilon + \phi.$$

This specification is a special case of a regression model with a general parameterized error covariance, of the form:

$$\Omega = \sigma^2(B'B)^{-1} + V$$

where σ^2 is the variance of ϕ, B is as before, and V is a diagonal matrix with as elements the heteroskedastic $x_i'\Sigma x_i$. The inverse Ω^{-1}, needed in ML or EGLS estimation is:

$$\Omega^{-1} = (1/\sigma^2).B'B.[(1/\sigma^2)B'B + V^{-1}]^{-1}.V^{-1}$$

Although complex and highly nonlinear, estimation and inference in this model can be carried out using the general principles outlined in Magnus (1978), Breusch and Pagan (1980), and Rothenberg (1984b).

9.4.2. Error Component Models for Cross Section Data

In many situations where observations over time and across space are combined (panel data), the regression error term can reasonably be decomposed into a spatial component, a time—specific componenent and an overall component. Formally,

$$\epsilon_{it} = \mu_i + \nu_t + \omega_{it}$$

where μ_i is the error component associated with spatial unit i, ν_t is associated with time t, and ω_{it} affects all observations equally. This model is treated in more detail in Section 10.2. Here, I consider a special form, which was suggested in Arora and Brown (1977) as an alternative specification of spatial autocorrelation.

In an cross—section, all errors by default pertain to spatial units, and the above decomposition simplifies to:

$$\epsilon_i = \mu_i + \omega$$

where μ_i is the individual—specific component, and ω affects all observations equally. The overall variance σ_e^2 is made up of a component associated with μ_i, σ_i^2 and a component associated with ω, σ_w^2. Since the latter is present in each cross—sectional covariance $E[\epsilon_i \epsilon_j]$, a special form of spatial autocorrelation results, with equal values for all observational pairs. In matrix notation, this special error covariance can be expressed as:

$$\Omega = \sigma^2.(I + \rho W)$$

where σ^2 is the combined variance of the two components, i.e., $\sigma_i^2 + \sigma_w^2$, ρ is the ratio of the variance σ_w^2 to σ^2, and W is a matrix of ones, except for zero elements on the diagonal.

Although this expression is similar to the equi—correlated error model in panel data, the lack of observations over the time dimension precludes its operational implementation. Indeed, the effect of the overall error component ω cannot be separated from the intercept in the model and thus cannot be identified. Consequently, this specification is of little practical use. It can therefore hardly qualify as an "improvement over current practice," as advertised in Arora and Brown.

9.4.3. Spatially Switching Regressions

Often, spatial heterogeneity is of a form that can be categorized into a small number of regimes, each represented by different values for the regression coefficients and/or different explanatory variables. Estimation in this situation can be carried out according to switching regression methods, originally developed by Quandt (1958).

When the number of different regimes and the switching points are known, maximum likelihood estimation can be carried out in a fairly straightforward way. Complications arise when the switching points are unknown, or when additional exogenous variables or stochastic functions are introduced to determine the classification into regimes.

In spatial modeling, an interesting aspect is the extent of spatial dependence that can be incorporated in combination with the different regimes. Although this has been extensively studied in a time series context, e.g., in Goldfeld and Quandt (1973) and Quandt (1981), a spatial analogue has not been developed. To fill this gap, I will briefly outline a number of ways in which spatial dependence can be incorporated.

For ease of exposition, consider a simple two−regime model, where the observations can be classified a priori into one or the other. Formally, the model can be expressed in the usual notation, with i and j as the respective regimes: [20]

$$
\begin{bmatrix} y_i \\ y_j \end{bmatrix} = \begin{bmatrix} X_i & 0 \\ 0 & X_j \end{bmatrix} \begin{bmatrix} \beta_i \\ \beta_j \end{bmatrix} + \begin{bmatrix} \mu_i \\ \mu_j \end{bmatrix}
$$

with the observations rearranged in accordance with the regimes. Keeping this special structure in mind, notation can be simplified by combining the variable and coefficient matrices for both subgroups as:

$$
y^* = X^* \beta^* + \mu^*
$$

where y^*, X^*, β^*, and μ^* correspond to the structure above.

In the standard Quandt approach, estimation is based on a likelihood for the overall model, in which the error terms have a different variance in each subset. The corresponding log likelihood function, conditional on the regime classification, is of the form (ignoring the usual constant):

$$
L = - (N_i/2).\ln(\sigma_i)^2 - (N_j/2).\ln(\sigma_j)^2 - (1/2).(\sigma_i)^{-2}.(y_i - X_i\beta_i)'(y_i - X_i\beta_i)
$$

$$
- (1/2).(\sigma_j)^{-2}.(y_j - X_j\beta_j)'(y_j - X_j\beta_j).
$$

Spatially autocorrelated error terms can be introduced in a number of ways, each of which results in a special case of the general specification introduced in Chapter 4. Consequently, maximum likelihood estimation and testing can be carried out following the same general principles as outlined in Section 6.2.

For example, one potential form for the spatial dependence would assume the same autoregressive process to affect all errors, as:

$$\mu^* = \lambda W \mu^* + \epsilon$$

where W is an N by N spatial weight matrix, and

$$E[\epsilon.\epsilon'] = \begin{bmatrix} (\sigma_i)^2.I_i & 0 \\ 0 & (\sigma_j)^2.I_j \end{bmatrix}$$

with I_i and I_j as identity matrices of dimensions N_i and N_j. The corresponding likelihood function is of the form:

$$L = - (N_i/2).\ln(\sigma_i)^2 - (N_j/2).\ln(\sigma_j)^2 + \ln|I - \lambda W| - (1/2)v'v$$

with

$$v'v = (y^* - X^*\beta^*)'(I - \lambda W)'\Omega^{-1}(I - \lambda W)(y^* - X^*\beta^*)$$

and

$$\Omega^{-1} = \begin{bmatrix} (\sigma_i)^{-2}.I_i & 0 \\ 0 & (\sigma_j)^{-2}.I_j \end{bmatrix}$$

Alternatively, a different process can be taken for each subgroup, with independence between the two regimes:

$$\mu^* = \begin{bmatrix} \lambda_i W_i & 0 \\ 0 & \lambda_j W_j \end{bmatrix} \begin{bmatrix} \mu_i \\ \mu_j \end{bmatrix} + \begin{bmatrix} \epsilon_i \\ \epsilon_j \end{bmatrix}$$

with $E[\epsilon.\epsilon']$ as before. The corresponding likelihood is obtained in a straightforward way, analogous to the previous case.

In the more complex situations of unspecified regimes or endogenous switching, inference is carried out within the framework of models with limited dependent variables and censored distributions. Since this is beyond the scope of this book, it will not be further considered here.

9.4.4. Spatial Adaptive Filtering

A completely different approach to dealing with spatial heterogeneity in regression coefficients is based on heuristic principles of adaptive estimation. In the spatial adaptive filtering technique (SAF), suggested by Foster and Gorr (1983, 1984, 1986), each observation has its unique set of coefficients, which are estimated and adjusted iteratively until a criterion of model fit is optimized. The SAF procedure is an extension to the spatial domain of adaptive estimation methods in time series analysis and forecasting, based on exponential smoothing and negative feedback. [21]

At each iteration of the adaptive process, the estimates for the coefficients at a location i are adjusted in function of an average of the estimates at neighboring locations j, and a criterion of predictive performance. Formally, the effect of each neighbor j is expressed as:

$$b_{kij}(l) = b_{kj}(l-1) + |b_{kj}(l-1)| \cdot \{(y_i - Y_{ij})/|Y_{ij}|\} \cdot \mu_k$$

where l is the iteration, k is the k-th regression coefficient, μ_k is a damping factor, and, for each i:

$$b_{ki}(l) = [1/N(i)] \cdot \Sigma_j \, b_{kij}(l)$$

is a spatial average of the estimates for locations j that are contiguous to i (b_{kij}), with N(i) as the number of neighbors for each i. Also, Y_{ij} is the predicted value for y at location i, determined by the estimates for the neighboring location j, in combination with the explanatory variables for i:

$$Y_{ij}(l) = \Sigma_k \, b_{kj}(l-1) \cdot x_{ki}.$$

The iterations are continued until a criterion of overall fit, such as the mean absolute percentage error (MAPE) is optimized.

Although the SAF approach provides fully spatially differentiated coefficient estimates, and may improve model fit (in-sample forecast) in applications, its statistical interpretation is limited. Indeed, the adaptive technique is an extreme example of the incidental parameter problem, in that each observation has its own vector of coefficients. Consequently, no probabilistic statements are possible, and the interpretation of the estimates is restricted to the peculiarities of the sample. The resulting lack of a framework for significance and other hypothesis testing limits the scope of the SAF technique primarily to exploratory analysis.

NOTES ON CHAPTER 9

[1]   Overviews of econometric issues related to the instability of parameters are given in, e.g., Swamy (1971, 1974), Belsley and Kuh (1973), Cooley and Prescott (1973, 1976), Rosenberg (1973), Pagan (1980), Raj and Ullah (1981), and Chow (1984). Instability of the functional form is discussed in, e.g., Quandt (1958, 1972, 1982), Goldfeld and Quandt (1973, 1976), Kiefer (1978), Quandt and Ramsey (1978), and Maddala (1983).

[2]   The tests in question are discussed in Glejser (1969), and Goldfeld and Quandt (1972). Other, more recent tests for heteroskedasticity are treated in, e.g., Harvey (1976), Godfrey (1978), Breusch and Pagan (1979), White (1980), Koenker and Bassett (1982), Cragg (1983), and MacKinnon and White (1985).

[3]   Early discussion of the Chow test and related issues of coefficient stability can be found in, e.g., Chow (1960), Zellner (1962), Fisher (1970), and Brown, Durbin and Evans (1975). The effect of heteroskedasticity on Chow—type tests has received considerably more attention in the literature than that of serial correlation, e.g., in Toyoda (1974), Jayatissa (1977), Schmidt and Sickles (1977), Watt (1979), and Honda (1982). See also Dufour (1982) for a recent review.

[4]   The simulation experiments consider a wide range of interactions between the spatial autocorrelation and extent of heteroskedasticity, with the latter varying from a relative error variance ratio of 1 (no heteroskedasticity) to 4. Based on the invariance results of Breusch (1980), they may therefore be considered quite general, since the effect of the regression coefficients can be ignored. The only other factor which needs to be further considered is the structure of the weight matrix. For additional details, see Anselin (1987b), and also Anselin and Griffith (1988).

[5]   See, e.g., Savin (1980) for a more extensive discussion.

[6]   Asymptotically, this is equivalent to a Wald or Likelihood Ratio test. However, since these necessitate the derivation of the model under the alternative hypothesis, they are considerably more complex computationally. As pointed out before, the Lagrange Multiplier is based on the model estimated under the null hypothesis, i.e., without any heteroskedasticity.

[7]   The block—diagonality of the information matrix pertains to the β coefficients and the parameters of the error covariance as a group, but not to the latter individually.

[8]   For a detailed derivation, see Anselin (1987b).

[9]   This is not a rigorous statement, since the residuals and error variance will differ between the two cases.

[10]   The LM, W and LR statistics are of this same form and differ only in whether the estimates for Ω and σ^2 are computed under the null hypothesis, the alternative or both.

[11]   Other, more complex situations for the error covariance structure could be considered as well. For example, the error variance could be heteroskedastic, or a different weight matrix may drive the spatial dependence in each subset of the data. The results for

136

these more complex cases can be found as direct extensions of the simple model considered here.

[12] See also Anselin (1987a), and Chapters 13 and 14. Recent overviews of the econometric problems associated with ad hoc specification searches are given in, e.g., Leamer (1974, 1978, 1983), Zellner (1979), Hendry (1980), Mayer (1980), Sims (1980), Frisch (1981), Malinvaud (1981), Lovell (1983), Ziemer (1984), and Cooley and LeRoy (1985, 1986).

[13] The orthogonal expansion approach, as used in, e.g., Casetti and Jones (1987, 1988) provides a means to avoid some of the multicolinearity problems. The original trend surface polynomial is replaced by a small number of principal components, which significantly reduces the number of expansion variables in the terminal model.

[14] For a similar approach in a time series context, see Singh et al. (1976).

[15] For example, if the trend is quadratic, two of the heteroskedastic components would be fourth powers of the coordinates. Since the coordinates in an infinitely large lattice approach infinity themselves, this is a situation which does not fit within the regularity conditions for an asymptotic approach, such as a bounded error variance. A similar problem is the potential infinity of the cross products of the explanatory variables. Although this does not fit within these regularity conditions either, most asymptotic results still hold if the so-called Grenander conditions are satisfied. For details, see, e.g., Judge et al. (1985, pp. 161-163) and also White (1984).

[16] For example, see the discussion in Grether and Maddala (1973), Granger and Newbold (1974), McCallum (1976), Thursby (1981), and Kiviet (1986).

[17] By convention, the regression equation is formulated without a disturbance term, since the error associated with the random intercept fullfils this role.

[18] For details, see, e.g., Swamy (1971, 1974), Magnus (1978), Raj and Ullah (1981), Schwallie (1982), and Hsiao (1986).

[19] This is in contrast to the statement by Arora and Brown (1977, p. 76), that the "random coefficient regression model overcomes the problem of spatial autocorrelation by assuming that the coefficients are random."

[20] In most applications of the switching regression approach in urban and regional analysis, the classification of observations into one or the other regime is done in analogy to the time series case. For example, in urban density studies, the observations are ordered with respect to distance from the CBD, and the switching points correspond to given distances (or distances to be determined by a search over all possible switching points). In general however, any organization into subsets of observations would be a valid classification into regimes.

[21] For an overview of the salient features of this technique in a time series context, see, e.g., Carbone and Longini (1977), Carbone and Gorr (1978), and Bretschneider and Gorr (1981, 1983).

CHAPTER 10

MODELS IN SPACE AND TIME

Up to this point in the book, the empirical context for the various estimators and tests has been limited to a purely cross—sectional situation. In this chapter, I consider models for which observations are available in two dimensions. Typically, one dimension pertains to space and the other to time, although other combinations, such as cross—sections of cross—sections and time series of time series can be encompassed as well. This situation has become increasingly relevant in a wide range of empirical contexts. It is referred to in the literature as panel data, longitudinal data, or pooled cross—section and time series data.

The idea of combining data for cross—sections and time series goes back to a suggestion of Marschak (1939), and has received considerable attention in the econometric literature. Recent overviews of the salient issues and suggested solutions can be found in Dielman (1983), Chamberlain (1984), Baltagi and Griffin (1984), and Hsiao (1985, 1986), among others.

The relevant literature is extensive and cannot be fully reflected within the more limited scope of this chapter. Since many estimation and testing issues that pertain to space—time data can be tackled by means of standard methods, their discussion in the current context would not be meaningful. To avoid duplication, these more familiar space—time methods will not be further considered. Also, as pointed out earlier, in Chapters 2 and 4, the extension of time series analysis to the space—time domain, as exemplified in the STARIMA modeling approaches of, e.g., Bennett (1979), Pfeiffer and Deutsch (1980a, 1980b), and Bronars and Jansen (1987), is outside the current scope.

In line with the general spirit of the book, the focus in this chapter will be on the complications that arise when spatial effects are present in models for space— time data. This issue is typically not considered in the standard econometric literature and therefore merits attention from a spatial econometric perspective. Specifically, I outline the implications of the presence of spatial dependence in two familiar types of space—time models: Seemingly Unrelated Regression (SUR) models, and Error Component models (ECM). In addition, I briefly discuss in general terms some aspects of spatial effects in simultaneous models.

10.1. SUR and Spatial SUR

The Seemingly Unrelated Regression (SUR) model, originally suggested by Zellner (1962) is designed for empirical situations where a limited degree of simultaneity is present in the form of dependence between the errors in different equations. If the equations pertain to time series for different regions, the resulting dependence can be considered as a form of spatial autocorrelation. This differs from the previously discussed spatial processes in that the spatial dependence is not expressed in terms of a particular parameterized function, but left unspecified as a general covariance.

In spatial econometrics, this model has been suggested as an alternative to the use of spatial weights, e.g., in Arora and Brown (1977). In addition, the SUR framework has been combined with more complex forms of spatial and time−wise dependence, e.g., in Hordijk and Nijkamp (1977, 1978), Hordijk (1979), Anselin (1980), Fik (1988), and applied to multiregional modeling, e.g., in White and Hewings (1982).

In this section, I will focus on the spatial form of the SUR model (spatial SUR), in which the regression equations pertain to cross−sections at different points in time, or to cross−sections of cross−sections. After a brief formal introduction of the model, I discuss estimation in two special situations. One is the spatial SUR model with spatial autocorrelation in the errors for each equation. The other is the case where spatially lagged dependent variables are included. Next, I outline a test for spatial residual autocorrelation in SUR models. The section is concluded with some general remarks on approaches to modeling nested spatial effects.

10.1.1. General Formulation

The SUR and spatial SUR are special cases in the general taxonomy of space−time models outlined in Chapter 4. In these models, the data on the dependent variable, y_{it}, and on vectors (1 by K) of explanatory variables, x_{it}, are organized by spatial units i (i=1,...,N) and time periods t (t=1,...,T).

In the most familiar SUR design, the regression coefficients β_i vary by spatial unit, but are constant over time. The error terms are spatially (contemporaneously) correlated, i.e., there is a constant covariance between errors for different spatial units at the same point in time. More formally, the model is expressed as in equation (4.11):

$$y_{it} = x_{it}\beta_i + \epsilon_{it}$$

with: [1]

$$E[\epsilon_{it} \cdot \epsilon_{jt}] = \sigma_{ij}.$$

In matrix form, the equation for each spatial unit i becomes:

$$y_i = X_i\beta_i + \epsilon_i$$

where y_i and ϵ_i are T by 1 vectors and X_i is a T by K_i matrix of explanatory variables. In general, the number of explanatory variables, K_i, can be different for each equation (spatial unit).

In the spatial SUR model, the coefficients β_t are constant across space, but vary for each time period. The error terms are temporally correlated, i.e., there is a constant covariance between errors for different time periods for the same spatial unit. More formally, the model is, as in equation (4.12):

$$y_{it} = x_{it}\beta_t + \epsilon_{it}$$

with

$$E[\epsilon_{it}\cdot\epsilon_{is}] = \sigma_{ts}.$$

In matrix form, the equation for each time period t becomes:

$$y_t = X_t\beta_t + \epsilon_t$$

where y_t and ϵ_t are N by 1 vectors and X_t is a N by K_t matrix of explanatory variables. As before, the number of explanatory variables, K_t, can be different for each equation (time period).

The spatial SUR model can be made operational only when more observations are available in the spatial dimension than in the time dimension (N>T). This is particularly suited for the situation where cross–sections are obtainable for a small number of time periods, as in a decennial census. In the more typical case where T>N, the usual SUR model applies.

Estimation and hypothesis testing in the SUR and spatial SUR model can be carried out as a special case of the framework with a general nonspherical error variance matrix. This is easiest illustrated when the regression equations are combined in stacked form. For the spatial SUR, the equations for time periods 1 to T are combined as:

$$\begin{bmatrix} y_1 \\ y_2 \\ y_T \end{bmatrix} = \begin{bmatrix} X_1 & 0 & \dots & 0 \\ 0 & X_2 & \dots & 0 \\ 0 & 0 & \dots & X_T \end{bmatrix} \begin{bmatrix} \beta_1 \\ \beta_2 \\ \beta_T \end{bmatrix} + \begin{bmatrix} \epsilon_1 \\ \epsilon_2 \\ \epsilon_T \end{bmatrix}$$

or, grouped,

$$Y = X\beta + \epsilon$$

where Y is a NT by 1 vector of dependent variables, K is the total number of coefficients $(=\Sigma_t K_t)$, X is a block diagonal matrix of dimensions NT by K, β is the overal coefficient vector of dimension K by 1, and ϵ is a NT by 1 error vector.

The dependence among error vectors is such that for each pair of time periods t,s:

$$E[\epsilon_t\cdot\epsilon_s'] = \sigma_{ts}.I.$$

This yields an overall error variance matrix Ω of the form:

$$E[\epsilon.\epsilon'] = \Omega = \Sigma \otimes I$$

where Σ is a T by T matrix with σ_{ts} as its elements, and $\otimes$ is the Kronecker product. [2]

When the elements of Σ are considered to be known, the usual generalized least squares (GLS) estimator can be applied to the full system, as:

$$b_{GLS} = [X'(\Sigma^{-1} \otimes I)X]^{-1}X'(\Sigma^{-1} \otimes I)y$$

with as covariance matrix:

$$var(b_{GLS}) = [X'(\Sigma^{-1} \otimes I)X]^{-1}.$$

The special structure of the error variance matrix Ω (and the use of Kronecker products) precludes that estimation would require inversion of matrices of the full NT dimension. Instead, only inverses of matrices of order K (for $X'\Omega^{-1}X$) and T (for Σ) are needed.

Typically, the elements of Σ are not known exactly, but need to be estimated together with the other model coefficients. As a consequence, inference is to be based on asymptotic considerations only, which is often ignored in applied work. As in Section 8.2.1, the proper approach consists of estimated generalized least squares (EGLS) or maximum likelihood (ML). Several iterative procedures have been suggested, e.g., as overviewed in Srivastava and Dwivedi (1979). Most approaches yield consistent (as well as unbiased) and asymptotically normal estimates, which form the basis for asymptotic significance tests. Finite sample properties are complex for all but the two–equation model. [3]

Two aspects of the SUR specification merit particular attention for modeling in applied regional science. One pertains to the issue of regional homogeneity and spatial aggregation, the other to testing for spatial dependence of an unspecified form.

The regional homogeneity problem deals with coefficient stability across regions, e.g., to assess the extent to which all regions in a spatial system respond to a given policy in the same way. In a SUR framework the behavior of each region can be modeled as a separate equation, related to the rest of the system by error covariance. A test for regional coefficient homogeneity can then be carried out in a straightforward way, as a hypothesis test on the equality of parameters in the SUR model. For example, if the interest focuses on a coefficient β_k, the hypothesis would take the form H_0: $\beta_{ki} = \beta_k$, $\forall$ i in the system. Similar hypotheses can be formulated for all or subsets of the model coefficients. If the joint null hypothesis of equality for all parameters in the model cannot be rejected, the data for all regions can reasonably be pooled. In a more general context, this is an important result with respect to aggregation, and one of the main motivations for the original Zellner (1962) paper.

As pointed out earlier, when the inter–equation covariances are not known exactly, inference in the SUR model (as a special case of a general nonspherical error variance) has only asymptotic validity. In the regional science literature on regional homogeneity, this important limitation is largely ignored. For example, two recently suggested F–tests for regional homogeneity, in Lin (1985) and Schulze (1987), are presented as exact tests, whereas they are really approximations to asymptotic results. As discussed in Section 6.3.5, inference in this case is not based on exact finite sample confidence levels (type I errors), but on adjustments of Wald, Likelihood Ratio or Lagrange Multiplier tests. Although these are asymptotically equivalent, they may lead to conflicting interpretation in finite samples. Exact results are unavailable, except for the simple two—equation case, e.g., in Smith and Choi (1982).

When the different equations in a SUR system pertain to regions, a test on spatial autocorrelation of an unspecified form is equivalent to a test on the diagonality of the inter—equation error covariance matrix Σ. In other words, a null hypothesis of no spatial dependence between the spatial units can be expressed as H_0: $\sigma_{ij}=0$, $\forall$ i,j with i $\neq$ j, i.e., a total of $1/2[N.(N-1)]$ constraints. Again, the tests are asymptotic in nature, and can be based on the Wald, LR or LM approach, e.g., as in Breusch and Pagan (1980). [4]

Several complications of the standard SUR model with T>N have been discussed in the literature. For example, serial autocorrelation in the intra—equation errors is analyzed in Parks (1967), Guilkey and Schmidt (1973), Maeshiro (1980), and Doran and Griffiths (1983). Heteroskedasticity is incorporated in Verbon (1980a) and Duncan (1983). The presence of lagged dependent variables, in combination with serial error autocorrelation is treated in Spencer (1979) and Wang, Hidiroglou and Fuller (1983). In addition, combinations of SUR with other modeling frameworks have been introduced, such as random coefficient models, in Singh and Ullah (1974), and error component models (ECM), in Avery (1977), Verbon (1980b), Baltagi (1980), and Prucha (1984). A detailed discussion of these issues is beyond the current scope.

10.1.2. Spatial SUR with Spatial Error Autocorrelation

Since the spatial SUR model consists of an equation for each time period, which is estimated for a cross—section of spatial units, intra—equation spatial error autocorrelation is potentially a problem. This is the spatial analogue of serial autocorrelation in the intra—equation disturbances. More formally, for each equation in the system:

$$y_t = X_t\beta_t + \epsilon_t$$

with

$$\epsilon_t = \lambda_t W\epsilon_t + \mu_t$$

and

$$E[\mu_t.\mu_s'] = \sigma_{ts}.I$$

In other words, the errors follow a spatial autoregressive process within each equation (with a different parameter for each t), as well as being correlated between equations. [5]

As before, the spatially dependent error vector ϵ_t can be considered as a transformation of the independent μ_t, as:

$$\epsilon_t = (I - \lambda_t W)^{-1}\mu_t$$

Consequently, it follows that:

$$E[\epsilon_t \cdot \epsilon_s{}'] \;=\; E[(I - \lambda_t W)^{-1} \mu_t \cdot \mu_s{}' (I - \lambda_s W')^{-1}]$$

$$\;=\; \sigma_{ts} \cdot [(I - \lambda_t W)'(I - \lambda_s W)]^{-1}$$

or,

$$E[\epsilon_t \cdot \epsilon_s{}'] \;=\; \sigma_{ts} \cdot B_t \cdot B_s{}'$$

where, for notational simplicity, [6]

$$B_t \;=\; (I - \lambda_t W)^{-1}.$$

The error covariance for the full system, Ω, takes the form:

$$\Omega \;=\; E[\epsilon \cdot \epsilon'] \;=\; B(\Sigma \otimes I)B' \tag{10.1}$$

where ϵ is a stacked NT by 1 error vector, Σ is the T by T inter–equation covariance matrix, and B is a blockdiagonal NT by NT matrix:

$$B \;=\; \begin{bmatrix} B_1 & 0 & \ldots & 0 \\ 0 & B_2 & \ldots & 0 \\ 0 & 0 & \ldots & B_T \end{bmatrix}$$

Alternatively, using the simplifying assumption of a constant W across equations, B can be expressed as:

$$B \;=\; [I - (\Lambda \otimes W)]^{-1} \tag{10.2}$$

where Λ is a T by T diagonal matrix containing λ_t, and I is a NT by NT identity matrix.

Estimation of this model is similar to the situation with time series data, although two main differences should be noted. As pointed out in Section 8.2, the usual iterative procedures developed for the timewise serial correlation case are no longer valid in the spatial model. Indeed, no simple estimate is available for the autoregressive parameters λ_t, so that an explicit nonlinear ML optimization is necessary. Although simple iterative analogues to the time case have been suggested, they turn out to be invalid, as illustrated in more detail below. On the other hand, the starting value problem of the time series approach is avoided, since all observations are used in the derivation of the ML result. However, boundary conditions may be more complex, in that they are present in two dimensions. This is further considered in Section 11.2.

Maximum likelihood estimates can be derived as a special case of a model with a general nonspherical error variance matrix, parameterized in terms of a small number of coefficients. [7] Under the assumption of normality, the log–likelihood function (ignoring constants) for the model expressed as stacked equations is of the general form:

$$L = - (1/2)\ln|\Omega| - (1/2)(Y-X\beta)'\Omega^{-1}(Y-X\beta)$$

with

$$|\Omega| = |B(\Sigma \otimes I)B'|$$

$$= |\Sigma|^N.|B|^2$$

based on the properties of determinants of Kronecker products. Consequently, the first term in the log−likelihood can be expressed as:

$$-(1/2)\ln|\Omega| = -(N/2)\ln|\Sigma| - \ln|B|$$

or, taking into account the block−diagonal structure and the presence of matrix inverses in B, as shown in (10.2):

$$-(1/2)\ln|\Omega| = -(N/2)\ln|\Sigma| + \Sigma_t \ln|I-\lambda_t W|$$

so that the complete log−likelihood becomes:

$$L = -(N/2)\ln|\Sigma| + \Sigma_t \ln|I-\lambda_t W|$$

$$-(1/2)(Y-X\beta)'[I-(\Lambda \otimes W')][\Sigma^{-1} \otimes I][I-(\Lambda \otimes W)](Y-X\beta).$$

The error variance Ω is a function of parameter vectors λ and σ, where λ is of dimension T by 1 (as many λ_t as equations or time periods), and σ is of dimension $(1/2)[T(T+1)]$ by 1 (the upper triangular elements of Σ). A complete derivation of the ML results is given in Appendix 10.A. The resulting first order conditions for the parameter estimates are, for the regression coefficients and inter−equation covariance:

$$\beta = \{X'[I-(\Lambda \otimes W')](\Sigma^{-1} \otimes I)[I-(\Lambda \otimes W)]X\}^{-1}$$

$$x \ X'[I-(\Lambda \otimes W')](\Sigma^{-1} \otimes I)[I-(\Lambda \otimes W)]y \qquad (10.3)$$

$$\Sigma = (1/N)Z'Z \qquad (10.4)$$

where Z is a N by T matrix of transformed residuals:

$$Z = [z_1, z_2,...,z_T] \qquad (10.5)$$

with

$$z_t = (I-\lambda_t W)e_t \qquad (10.6)$$

$$= e_t - \lambda_t We_t$$

and

$$e_t = y_t - X_t\beta_t. \qquad (10.7)$$

The estimates for the λ_t are obtained as the solution to T nonlinear equations, for each t of the form:

$$\text{tr } W(I-\lambda_t W)^{-1} = \Sigma_h \, \sigma^{\text{th}}.[e_t\text{'}W\text{'}(I-\lambda_h W)e_h] \tag{10.8}$$

where σ^{th} is the t,h element in the inverse matrix Σ^{-1}. Expression (10.8) clearly illustrates that the usual iterative procedure, which derives an estimate for λ_t in isolation from the values for the other λ_h in the system will lead to results that are not maximum likelihood. Only when the σth are zero $\forall$ h $\neq$ t, i.e., when there is no inter–equation covariance, does (10.8) reduce to the familiar ML conditions for spatially correlated disturbances.

The first order conditions (10.3), (10.4), and (10.8), and definitions (10.6) and (10.7) form a highly nonlinear system of equations, which needs to be solved for all parameters β, λ, and σ. [8] Clearly, although a simultaneous solution is possible by means of numerical optimization techniques, it would be computationally cumbersome. An alternative is an iterative approach, in which at each iteration the values for λ_t are solved conditional upon results for β (leading to residuals e_t) and Σ^{-1} from the previous iteration. Many forms of this iterative GLS are possible, and most lead to convergent local maxima, as shown in Oberhoffer and Kmenta (1974) for a general GLS model. One alternative would be to start with OLS in each equation, derive an estimate for Σ from the OLS residuals, and use these results to solve the T equations for λ_t. In subsequent rounds, the values for λ_t and Σ are used in the EGLS estimation for a new set of β. This leads to a sequence of iterations until a convergence criterion of precision has been satisfied. Alternatively, the OLS residuals could be used to estimate λ_t (ignoring inter–equation covariance) in a first round. Subsequently, ML estimates for β (using the λ_t) and a new set of residuals would lead to an estimate for Σ^{-1}. This is then used in the next iteration to obtain new estimates for β and λ_t. A more detailed discussion of these operational issues is presented in Section 12.1.

The information matrix for the coefficients is block diagonal between the elements of β and of [λ, σ], as shown for the general nonspherical model by Breusch (1980). However, in contrast to the SUR result with serial autocorrelation in the time dimension, given in Magnus (1978), the matrix is not block diagonal between the elements of λ and σ, due to the two–directional nature of dependence in space. [9]

The relevant partitions of the information matrix yield, for the different combinations of coefficients:

– for the elements of β:

$$I(\beta,\beta) = X\text{'}[I-(\Lambda \otimes W\text{'})](\Sigma^{-1} \otimes I)[I-(\Lambda \otimes W)]X$$

– for the elements of λ, with $D_t=W(I-\lambda_t W)^{-1}$:

$$I(\lambda_t,\lambda_t) = \text{tr } (D_t)^2 + \sigma^{\text{tt}}.\sigma_{tt}.\text{tr } D_t\text{'}D_t$$

$$I(\lambda_t,\lambda_s) = \sigma^{\text{ts}}.\sigma_{ts}.\text{tr } D_t\text{'}D_s$$

where σ^{ts} and σ_{ts} are the t,s elements of Σ^{-1} and Σ respectively;

— for the elements of λ_t, σ_{hk}:

$$I(\lambda_t, \sigma_{hk}) \quad = \quad \text{tr } (E^{tt}\Sigma^{-1}E^{hk}).\text{tr } D_t$$

where E^{tt} is a T by T matrix of zeros, except for element t,t, which equals one, and E^{hk} is a T by T matrix of zeros, except for elements h,k and k,h, which equal one. The matrices E^{tt} and E^{hk} are used to select the relevant elements from the inverse Σ^{-1};

— for the elements of σ:

$$I(\sigma_{ij}, \sigma_{hk}) \quad = \quad (1/2).N.\{\text{tr } (\Sigma^{-1}E^{ij}).(\Sigma^{-1}E^{hk})\}$$

with the E matrices constructed as before.

The inverse of the information matrix yields an asymptotic variance matrix for the coefficients, which can be used to construct hypothesis tests by means of the Wald, LR or LM approaches. A complete derivation of the information matrix is presented in Appendix 10.B. A set of fully written out results for the simple two equation case is given in Appendix 10.C. An empirical illustration is presented in Section 12.3.6.

10.1.3. Spatial SUR with Spatially Lagged Dependent Variables

As pointed out in Section 6.1.1, the presence of a spatially lagged dependent variable in and of itself is sufficient to preclude OLS from yielding consistent estimates. This is the case irrespective of dependence or independence in the error terms. As a consequence, when spatially lagged dependent variables are present the initial step in a spatial SUR procedure cannot be based on OLS estimation.

This situation is similar to the one for disturbance related simultaneous equation models, and can be approached from either an ML or an IV perspective. Formally, each equation in the system is expressed as:

$$y_t = \gamma_t W y_t + X_t \beta_t + \epsilon_t$$

or,

$$A_t y_t = X_t \beta_t + \epsilon_t$$

with,

$$A_t = I - \gamma_t W$$

and, as before,

$$E[\epsilon_t . \epsilon_s'] = \sigma_{ts}.$$

The full system can be represented in stacked form as:

$$AY - X\beta = \epsilon$$

where,

$$A = I - (\Gamma \otimes W)$$

with Γ as a T by T diagonal matrix with γ_t on the diagonal, and I as an identity matrix of dimension NT. Similar to the approach taken for the general model in Section 6.2.3, a further transformation yields:

$$(\Sigma \otimes I)^{-1/2}(AY - X\beta) = \nu$$

where ν is a NT by 1 vector of independent standard normal disturbances.

Under the assumption of normality, the likelihood for this model is obtained as before, by utilizing the following Jacobian:

$$
\begin{aligned}
J &= |(\Sigma \otimes I)^{-1/2}.A| \\
&= |\Sigma|^{-N/2}.|A|.
\end{aligned}
$$

The corresponding logarithmic likelihood (ignoring the constant term) is:

$$L = -(N/2)\ln|\Sigma| + \ln|A| - (1/2)\nu'\nu$$

with

$$\nu'\nu = (AY - X\beta)'(\Sigma^{-1} \otimes I)(AY - X\beta)$$

or, using the block diagonal structure of A, L becomes:

$$L = -(N/2)\ln|\Sigma| + \Sigma_t \ln|I-\gamma_t W| - (1/2)\nu'\nu.$$

Maximum likelihood estimates for the parameters can be obtained in the usual fashion by numerical optimization. [10]

An alternative approach to estimation can be based on instrumental variables techniques, in particular the three stage least squares (3SLS) estimator. As in Section 7.1, the lagged dependent variable and the other explanatory variables can be grouped in a matrix $Z_t = [y_L, X]_t$, and expressed in stacked form as Z. With a matrix of instruments Q_t for each equation, stacked in Q, the IV estimator for a general nonspherical error variance $\Omega = \Sigma \otimes I$ can be applied as:

$$b_{IV} = \{Z'Q[Q'(\Sigma \otimes I)Q]^{-1}Q'Z\}^{-1}Z'Q[Q'(\Sigma \otimes I)Q]^{-1}Q'y$$

Alternatively, the same set of instruments can be used for all equations in the system, which results in the 3SLS estimator:

$$b_{3SLS} = \{Z'[\Sigma^{-1} \otimes Q(Q'Q)^{-1}Q']Z\}^{-1}Z'[\Sigma^{-1} \otimes Q(Q'Q)^{-1}Q']y$$

with as variance matrix:

$$\mathrm{var}(b_{3SLS}) = \{Z'[\Sigma^{-1} \otimes Q(Q'Q)^{-1}Q']Z\}^{-1}.$$

An operational implementation of this approach is based on estimates for the matrix Σ, and is carried out in a number of iterations. A simple three step approach would start by first estimating each equation separately, using two stage least squares (2SLS). The resulting residuals lead to an estimate for Σ^{-1}, which is used in the third step to obtain b_{3SLS}. Further iterations can be carried out until a convergence criterion is met. The coefficient estimates are asymptotically normally distributed with a covariance matrix as above. This forms the basis for asymptotic hypothesis tests using Wald, Likelihood Ratio or Lagrange Multiplier tests.

Clearly, compared to the nonlinear ML optimization, the 3SLS approach is easier to implement. However, the resulting estimates may yield explosive spatial autoregressive parameters, as discussed in Section 7.1. Often, the 3SLS estimates can be used as good starting values in the nonlinear ML optimization, and thereby considerably speed up the computations. The 3SLS approach is illustrated in Section 12.3.3.

10.1.4. Testing for Spatial Autocorrelation in Spatial SUR Models

Since the estimation of the spatial SUR model becomes considerably more complex when spatial dependence is present in the intra—equation disturbances, it is useful to test for this complication. Some extensions of the Moran I coefficient to space—time models were suggested by Martin and Oeppen (1975) and Hordijk and Nijkamp (1977). However, neither approach has known distributional properties, which limits their usefulness in a rigorous model specification framework. [11]

An alternative test, with attractive asymptotic properties, can be based on the Lagrange Multiplier principle. The null hypothesis of interest consists of T constraints, as H_0: $\lambda=0$, where, as before, λ is a T by 1 vector which contains the λ_t coefficients for each equation. Following the usual LM approach, the coefficient vector is partitioned as:

$$\theta = [\lambda \mid \sigma \ \beta]$$

where σ contains the upper triangular elements of Σ.

The test statistic is constructed by deriving the score vector and the relevant partitioned inverse of the information matrix under the null. Assuming normality, and with the same log—likelihood as in Section 10.1.2, the score is found to be, for each λ_t:

$$\partial L/\partial \lambda_t = -\mathrm{tr}W(I-\lambda_t W)^{-1} + \epsilon'[I_{NT}-(\Lambda \otimes W')(\Sigma^{-1} \otimes I_N)(E^{tt} \otimes W)]\epsilon$$

with ϵ as a NT by 1 error vector, Λ as a T by T diagonal matrix with elements λ_t, E^{tt} as a T by T matrix of zeros, except for a one in position t,t, and I_{NT} and I_N as identity matrices of the proper dimensions.

Under the null hypothesis, $\lambda_t=0$, and Λ becomes a zero matrix, which reduces the score to:

$$\partial L/\partial \lambda_t = - \text{ tr } W + \epsilon'(\Sigma^{-1} \otimes I)(E^{tt} \otimes W)\epsilon.$$

Since W has zero diagonal elements by convention, this expression further simplifies to:

$$\partial L/\partial \lambda_t \quad = \quad \epsilon'[(\Sigma^{-1}.E^{tt}) \otimes W]\epsilon$$

$$= \quad \Sigma_h \; \sigma^{ht}.\epsilon_h{}'W\epsilon_t.$$

The score vector (as a row vector) for all T λ_t can be expressed succintly as:

$$\iota'(\Sigma^{-1}*U'WU)$$

where ι is a T by 1 vector of ones, U is a N by T matrix with the errors as columns, and * stands for the Hadamard product. [12]

As pointed out in Section 10.1.2, the information matrix in this model is not block diagonal between the elements of λ and σ. At first sight, this seems to complicate the derivation of the partitioned inverse considerably. However, under the null:

$$\text{tr } D_t = \text{tr } W = 0$$

so that the inverse can be obtained from the elements $\Psi(\lambda,\lambda)$ that correspond to λ only. Under the null, these become:

$$\Psi(\lambda_t,\lambda_t) = \text{tr } W^2 + \sigma^{tt}.\sigma_{tt}.\text{tr } (W'W)$$

$$\Psi(\lambda_t,\lambda_s) = \sigma^{ts}.\sigma_{ts}.\text{tr } (W'W).$$

With the simplifying notation:

$$T_1 = \text{tr } W'W$$

$$T_2 = \text{tr } W^2$$

the matrix $\Psi(\lambda,\lambda)$ can be expressed succintly as:

$$\Psi(\lambda,\lambda) = T_2.I + T_1.\Sigma^{-1}*\Sigma.$$

As a result, the LM test statistic for spatial error autocorrelation in the spatial SUR model becomes:

$$LM = \iota'(\Sigma^{-1}*U'WU)[T_2.I + T_1.\Sigma^{-1}*\Sigma]^{-1}(\Sigma^{-1}*U'WU)'\iota$$

which is distributed asymptotically as a χ^2 with T degrees of freedom. This test can be constructed in a straightforward way from the results of a standard SUR, with the additional computation of two matrix traces and the spatially weighted cross products U'WU. An illustration of the expression for the two equation SUR model is presented in Appendix 10.D, and an empirical implementation is given in Section 12.3.4.

10.1.5. Nested Spatial Effects

The organization of data in two dimensions does not always correspond to a combination of time series and cross sections. For example, if employment models for different sectors are estimated for the same cross section of counties, the design consists of a cross section of cross sections. In this case, the error covariance between the different equations reflects a dependence between the different sectors. The structure of this dependence is such that it is constant for all counties. More formally, with sectors r and s and cross sectional elements i, j, the covariance is:

$$E[\epsilon_{ri} \cdot \epsilon_{sj}] = \sigma_{rs}, \; \forall \; i=j$$

$$E[\epsilon_{ri} \cdot \epsilon_{sj}] = 0, \; \text{for } i \neq j.$$

Since estimation is based on aggregate spatial units with arbitrary boundaries, spatial error autocorrelation could be present within each cross section. In this instance, the estimator in Section 10.1.2. and the test in Section 10.1.3. are particularly relevant.

Still a different situation occurs when spatial units are grouped in clusters, and estimation is carried out for a cross section in each cluster. If the errors between the elements in the two clusters show dependence, a particular form of the spatial SUR model results. [13] Moreover, if the errors within each cluster equation show spatial dependence, a certain form of nested spatial effects is present, in the sense of dependence within as well as between the clusters. However, this case does not necessarily conform to the SUR framework. Indeed, the elements in each cluster need to be matched to those in the other, such that each is related to only one in the other cluster. This follows from the requirement that $E[\epsilon_{ri} \cdot \epsilon_{sj}] = 0$ when $i \neq j$, in the same notation as above, with r and s representing the clusters. Clearly, this is a rather restrictive structure, with a limited scope for application. One example where it may be relevant would be in the study of two linear border regions, one in each country (or two linear sets of counties, separated by a state border or a river), with each spatial unit in one country exactly matched to one region in the other country.

In the more general situation where intra— as well as inter—cluster dependence is present, the model becomes a special case of a non—spherical error variance matrix. For example, in a simple two equation framework, the following types of spatial dependencies could be present:

$$\epsilon_i = \lambda_i W \epsilon_i + \mu_i, \; \text{for } i=1,2$$

and

$$E[\mu_{1h} \cdot \mu_{2k}] = \sigma_{12}, \; \forall \; h,k.$$

The second type of dependence implies a constant variance between the errors of the two clusters, for each pair of spatial units h,k. This is similar to the dependence implied by an error components model, as discussed in more detail in the next section. Formally, the error covariance for the two equation system becomes:

$$\Omega = [I - (\Lambda \otimes W)]^{-1} U [I - (\Lambda \otimes W')]^{-1}$$

with,

$$U = \begin{bmatrix} \sigma_{11}.I & \sigma_{12}.E \\ \sigma_{21}.E & \sigma_{22}.I \end{bmatrix}$$

where E is a N by N matrix of ones. As in the other spatial SUR models, estimation of the parameters in this case can be carried out as a special application of the general ML approach of Magnus (1978).

10.2. Error Component Models in Space–Time

An alternative way in which space and time effects can be incorporated into a regression model consists of the error components (ECM) or variance components (VCM) approach. In this framework, the space (or individual) and time effects are considered as part of the unobservable random error, which results in models with a nonspherical disturbance variance. In contrast, in the so–called fixed effects models, the space and time aspects are expressed by means of dummy variables.

The ECM structure has received widespread use in applied econometric analyses of panel data, going back to the well–known studies by Balestra and Nerlove (1966), Wallace and Hussain (1969), Nerlove (1971), and Maddala (1971). More recently, attention has focused on the incorporation of various forms of dynamics in time, e.g., in Lillard and Willis (1978), Lillard and Weiss (1979), Anderson and Hsiao (1981, 1982), MaCurdy (1982), Bhargava and Sargan (1983), as well as on the acknowledgement of different types of heteroskedasticity, e.g., in Mazodier and Trognon (1978), and Wansbeek and Kapteyn (1982). Various estimation issues are discussed in Chamberlain (1982), Magnus (1982), Amemiya and MaCurdy (1986), Hsiao (1986), and Breusch (1987).

In spatial econometrics, the ECM approach was suggested as an alternative way of dealing with spatial autocorrelation in Arora and Brown (1977). However, in general, spatial dependence in error component models has been largely ignored. In the remainder of this section, after a brief outline of the formal aspects of the ECM, I present an illustration of how spatial dependence can be incorporated and how tests for its presence can be developed.

10.2.1. General Formulation

As in the previous section, the data on the dependent variable y_{it} and the explanatory variables x_{it} are organized by spatial units i (i=1,...,N) and time periods t (t=1,...,T). However, in contrast to the SUR model, in the error component model the regression coefficients are considered fixed, as in:

$$y_{it} = x_{it}\beta + \epsilon_{it}.$$

The error term is assumed to incorporate unobserved effects due to space, μ_i, and due to time, λ_t, as well as the usual disturbance, ϕ_{it}. For example, the unobservable space—specific effects could be due to aspects of regional structure, and time—specific effects could be related to trends or cycles. The space—specific effect is assumed to be constant over time, while the time—specific effect is assumed to be constant across space. Formally, the overall error consists of three components, as:

$$\epsilon_{it} = \mu_i + \lambda_t + \phi_{it}$$

to form the three error components model (3ECM). The situation where either space or time—specific effects are included, but not both, is called the two error components model (2ECM).

In contrast to the fixed effects approach, where the space or time—specific aspects are expressed as a dummy variable, in the ECM these aspects are considered to be random. Typically, the three components have zero expected value and are assumed to be independently distributed. In addition, constraints are imposed to exclude explicit dependence across space or over time:

$$E[\mu_i.\mu_j] = 0 \text{ for } i \neq j, \text{ but } E[\mu_i.\mu_i] = \sigma(\mu), \forall i$$

$$E[\lambda_t.\lambda_s] = 0 \text{ for } t \neq s, \text{ but } E[\lambda_t.\lambda_t] = \sigma(\lambda), \forall t$$

The third error component, ϕ_{it}, is usually assumed to be independent and homoskedastic:

$$E[\phi_{it}.\phi_{js}] = 0 \text{ for } i \neq j, \text{ and } t \neq s, \text{ but}$$

$$E[(\phi_{it})^2] = \sigma, \forall i,t.$$

Even though no explicit error dependence is present, an implicit type of spatial correlation and temporal correlation results. Indeed, since the term μ_i is common to ϵ_{it} for all time periods, the covariance between ϵ_{it} and ϵ_{is} will contain an element $\sigma(\mu)$. Similarly, the presence of the term λ_t for all ϵ_{it} in the same period leads to a spatial covariance of $\sigma(\lambda)$ between ϵ_{it} and ϵ_{jt}. In other words, the spatial error component induces a form of temporal autocorrelation, whereas the time specific error component leads to a form of spatial autocorrelation. However, this type of autocorrelation is unusual, in that its magnitude is constant, irrespective of the time lag or distance between observations. In other words, the errors are equicorrelated across space as $\sigma(\lambda)$, and equicorrelated over time as $\sigma(\mu)$.

The ECM structure can be expressed succinctly in matrix notation, as:

$$y = X\beta + \epsilon$$

where y is a NT by 1 vector of observations on the dependent variable, X is a NT by K matrix of explanatory variables, β a K by 1 vector of coefficients, and ϵ a NT by 1 vector of errors.

The data can be grouped by spatial unit or by time period, which leads to a slightly different notation for the components of the error term. The more usual approach is to sort first by spatial unit, and then by time. For example, for the errors in the first group (i.e., for i=1) the overal term would consist of a sum of three T by 1 vectors, one with μ_1 (the same $\forall$ t), one with λ_t (different for each t) and one with ϕ_{it}, or:

$$\epsilon_1 = \iota_T \cdot \mu_1 + I_T \cdot \lambda + \phi_{it}$$

where ϵ_1 and ϕ_{it} are T by 1 vectors of errors, ι_T is a T by 1 vector of ones, I_T is an identity matrix of dimension T, λ is the T by 1 vector of time components, and μ_1 is the first space component.

The full notation, for N sets of T observations on each spatial unit over time becomes:

$$\epsilon = (I_N \otimes \iota_T)\mu + (\iota_N \otimes I_T)\lambda + \phi$$

with I as an identity matrix and ι as a vector of ones, both of the dimensions in the subscripts, and μ as a N by 1 vector of spatial components. Alternatively, when the observations are sorted by time, and represent T sets of N observations on each time period for all spatial units, the error structure takes on the form:

$$\epsilon = (\iota_T \otimes I_N)\mu + (I_T \otimes \iota_N)\lambda + \phi.$$

In each case, a particular nonspherical error variance results. For the first organization, by spatial unit, this takes the form:

$$\Omega = \sigma(\mu)[I_N \otimes \iota_T \cdot \iota_T'] + \sigma(\lambda)[\iota_N \cdot \iota_N' \otimes I_T] + \sigma.I$$

with the σ as the variances of the respective components. In the second organization of data, by time period, the error variance becomes:

$$\Omega = \sigma(\mu)[\iota_T \cdot \iota_T' \otimes I_N] + \sigma(\lambda)[I_T \otimes \iota_N \cdot \iota_N'] + \sigma.I.$$

The estimation of the parameters in the ECM structure can be carried out by EGLS or maximum likelihood. The special form for Ω leads to an inverse Ω^{-1} that can be expressed in terms of inverses of order T or N, so that the inversion of a full NT by NT matrix can be avoided. [14]

10.2.2. Spatial Autocorrelation in Error Component Models

The equicorrelated form of spatial dependence which is implied by the error component model does not allow for distance decay effects. Since this runs counter to accepted spatial interaction theory, it may not be a very useful structure in applied regional science. However, it is important to realize that the spatial equicorrelation is due to the time specific component, and not to the spatial component in the error. Different forms of dependence or heteroskedasticity can be introduced in the ECM expression by imposing the appropriate structure on the disturbance component ϕ_{it}. In this way, it becomes possible to combine a spatial error component μ_i with a spatial autoregressive dependence in the error ϕ_{it}. [15]

For example, in the 2ECM framework, with the data organized by time period, the error can be composed of a spatial effect μ_i and an overall disturbance ϕ_{it}. The latter can be taken to follow a spatial autoregressive process across spatial units in each time period. Formally, the error vector for time t can be presented as:

$$\epsilon_t = \mu + \phi_t$$

with

$$\phi_t = \lambda.W.\phi_t + \nu_t$$

where λ is a spatial autoregressive coefficient (and not, as before, the time effect), and W is a N by N weight matrix. As before,

$$\phi_t = (I-\lambda W)^{-1}\nu_t = B^{-1}\nu_t$$

with B=I−λW for notational simplicity. If the parameter λ is taken to be fixed for all time periods, the overall error structure becomes:

$$\epsilon = (\iota_T \otimes I_N)\mu + (I_T \otimes B^{-1})\nu$$

with the same notation as in the previous section.

The overall NT by NT error variance matrix Ω is obtained in a straightforward manner, with the component variances as:

$$E[(\mu_i)^2] = \sigma(\mu)$$
$$E[(\phi_{it})^2] = \sigma$$

and all other covariances equal to zero:

$$\Omega = \sigma(\mu).(\iota_T.\iota_T' \otimes I_N) + \sigma.[I_T \otimes (B'B)^{-1}]$$

or, with

$$\omega = \sigma(\mu)/\sigma$$
$$\Omega = \sigma.\Psi$$
$$\Psi = \iota_T.\iota_T' \otimes \omega.I_N + [I_T \otimes (B'B)^{-1}].$$

Estimation can be carried out by maximum likelihood. Under the assumption of normality, the general log likelihood (ignoring constants) is of the form:

$$L = -(NT/2)\ln \sigma - (1/2)\ln |\Psi| - (1/2\sigma)e'\Psi^{-1}e$$

with

$$e = y - X\beta.$$

Although, at first sight, the determinants and inverse matrices involved in these expressions seem quite formidable, their special structure allows for considerable simplification. After the application of a number of matrix properties from Magnus (1982), [16] the determinant can be shown to take the form:

$$|\Psi| = |(B'B)^{-1} + (\omega.T).I_N|.|B|^{-2(T-1)}$$

and the inverse becomes:

$$\Psi^{-1} = (1/T)\iota_T.\iota_T' \otimes [(B'B)^{-1} + (\omega.T).I_N]^{-1} + [I_N - (1/T)\iota_T.\iota_T'] \otimes B'B$$

which involves no matrix inversions of dimension larger than N.

Consequently, the log likelihood can be written as:

$$L = -(NT/2)\ln\sigma -(1/2)\ln|(B'B)^{-1} + (\omega T)I_N| -(T-1)\ln|B|$$

$$-(1/2\sigma)e'\{(1/T)\iota_T.\iota_T' \otimes [(B'B)^{-1} +(\omega T)I_N]^{-1}\}e$$

$$-(1/2\sigma)e'\{[I_N-(1/T)\iota_T.\iota_T'] \otimes B'B\}e$$

First order conditions for the ML estimates and the elements of the information matrix can be obtained in the usual way, as an application of the approach for a general parameterized error covariance.

10.2.3. A Lagrange Multiplier Test for Spatial Autocorrelation in Error Component Models

Similar to the approach taken in the SUR model, an asymptotic test for the presence of spatial autocorrelation in the error component model can be based on the Lagrange multiplier principle. The partitioning of the coefficient vector that corresponds to the null hypothesis H_0: $\lambda=0$ is:

$$[\lambda \mid \sigma \ \omega \ \beta]$$

in the same notation as in the previous section.

As shown in more detail in Appendix 10.E, the score vector for λ turns out to be of the form:

$$\partial L/\partial\lambda = \text{tr } \Psi^{-1}[I_T \otimes (B'B)^{-1}(W'B+B'W)(B'B)^{-1}]$$

$$+(1/2\sigma)e'\Psi^{-1}[I_T \otimes (B'B)^{-1}(W'B+B'W)(B'B)^{-1}]\Psi^{-1}e$$

Under the null hypothesis, i.e., with $\lambda=0$, the following simplifying results hold:

$$B = B'B = (B'B)^{-1} = I$$

$$\Psi = (\iota_T.\iota_T' \otimes \omega.I_N) + I_{NT}$$

Also, the error variance Ψ is the same as for the more familiar 2ECM structure (organized by time period), and its inverse reduces to: [17]

$$\Psi^{-1} = I_{NT} - \iota_T.\iota_T' \otimes (\omega/1+T\omega).I_N.$$

Consequently, the score for λ simplifies greatly under the null hypothesis, and becomes:

$$\partial L/\partial \lambda = (1/\sigma).e'\{[I_T + \kappa(T\kappa-2).\iota_T.\iota_T'] \otimes W\}e \qquad (10.9)$$

where $\kappa=\omega/1+T\omega$ for notational simplicity. This expression consists of cross products of spatially weighted residuals, similar to those found in the Moran coefficient, but adjusted for the variance components.

Although the information matrix is not block diagonal between λ and the σ and ω parameters, it achieves this property under the null, due to the zero traces of the matrices W and W', as shown in Appendix 10.E. Consequently, the partitioned inverse of the information matrix reduces to a simple inverse for the element corresponding to λ. In general, this is the inverse of the expression:

$$I(\lambda,\lambda) = (1/2)\ \mathrm{tr}\ \Psi^{-1}(\partial\Psi/\partial\lambda)\Psi^{-1}(\partial\Psi/\partial\lambda)$$

which, under the null simplifies to:

$$I(\lambda,\lambda) = (1/2)\ \mathrm{tr}\ [I_{NT} - E \otimes \kappa.I_N][I_T \otimes (W+W')]$$
$$\times[I_{NT} - E \otimes \kappa.I_N][I_T \otimes (W+W')]$$

with $E=\iota_T.\iota_T'$ for notational simplicity. After a straightforward application of matrix algebra, this expression becomes:

$$I(\lambda,\lambda) = (T^2\kappa^2 - 2\kappa + T).(T_1 + T_2)$$

with, as before,

$$T_1 = \mathrm{tr}\ W'W$$

$$T_2 = \mathrm{tr}\ W^2.$$

As a result, the LM statistic is found as:

$$LM = \{(1/\sigma)e'\{[I_T + \kappa(T\kappa-2).E] \otimes W\}e\}^2 / p \sim \chi^2(1)$$

with e as the residual vector, and,

$$p = (T^2\kappa^2 - 2\kappa + T).(T_1 + T_2)$$

or, equivalently:

$$(1/\sigma)e'\{[I_T + \kappa(T\kappa-2).E] \otimes W\}e / p^{1/2} \sim N(0,1).$$

This statistic can be computed from the output of a standard 2ECM program by constructing the appropriate spatially weighted cross products and calculating two matrix traces. An empirical illustration is presented in Section 12.3.5.

10.3. Simultaneous Models in Space−Time

The potential for modeling spatial dependence when observations are available over time and across space is not limited to error variance structures. Indeed, the additional dimension in the data allows for the explicit analysis of the interaction between spatial units in the form of a system of simultaneous equations.

For a fixed number of spatial units N, and with T>N, a phenomenon in each region can be expressed as a separate equation. In each of these equations, the dependent variable for one or more other regions can be included as an endogenous variable, either contemporaneously or lagged in time. Formally, the equation for each region can be expressed as:

$$y_i = Y\delta + X\beta + \epsilon$$

where Y is potentially a N−1 by T matrix of observations on the spatial endogenous variables, δ is a N−1 by 1 vector of spatial coefficients, X is a K by T matrix of other endogenous and exogenous variables, with coefficients β, and ϵ is the error term.

The full system of spatial equations can be expressed as: [18]

$$Y\Delta + XB = E$$

where the matrix of all spatial variables Y is N by T (N equations, one for each spatial unit), and Δ is a N by N matrix of spatial coefficients. Identification of this system is obtained in the usual fashion by imposing restrictions on the coefficient matrices and/or error covariances. For example, the requirement of symmetry could be imposed on the spatial coefficient matrix Δ.

This framework is the most general spatial modeling structure. It allows for explicit spatial dependence in the Δ coefficients as well as for dependence to be incorporated in the error terms, as in the SUR model. Estimation can be carried out by means of limited information or full information maximum likelihood, and 2SLS or 3SLS, in the usual manner.

Appendix 10.A.:   **First Order Conditions for the ML Estimator in the Spatial SUR Model with Spatially Autocorrelated Errors.**

As shown in Magnus (1978), the first order conditions for ML estimation of the parameters in a linear regression model with a general nonspherical error variance of the form $\Omega(\theta)$ are:

$$\beta = (X'\Omega^{-1}X)^{-1}X'\Omega^{-1}y$$

$$\text{tr}\{(\partial\Omega^{-1}/\partial\theta_i).\Omega\} = e'(\partial\Omega^{-1}/\partial\theta_i)e, \; \forall \; i$$

with

$$e = y - X\beta.$$

In situations where the matrix Ω takes on a special structure, the first order conditions can be expressed more explicitly. Specifically, for the SUR model with spatially dependent errors, the matrix Ω can be expressed as in (10.1):

$$\Omega = B(\Sigma \otimes I)B'$$

which is a special case of the model considered in Magnus (1978). [19] The conditions for β and Σ follow as a direct application and need no further special consideration.

The first order conditions for the parameters λ_t are found from:

$$\text{tr}\{(\partial B^{-1}/\partial\lambda_t).B\} = \text{tr}\; Z\Sigma^{-1}Z_t', \; \forall \; t \qquad (10.A.1)$$

where Z is the error transformation as in (10.5) and (10.6), and Z_t is an N by T matrix:

$$Z_t = [z_{1t}, z_{2t},..., z_{Tt}] \qquad (10.A.2)$$

with

$$z_{ht} = (\partial B_h^{-1}/\partial\lambda_t).e_h \qquad (10.A.3)$$

for h=1,...,T and t=1,...,T.

Since B^{-1} is a block diagonal matrix with elements $(I-\lambda_t W)$, its partial derivative with respect to a particular λ_t consists of $-W$ in the t,t block on the diagonal, and zeros elsewhere, or:

$$\partial B_h^{-1}/\partial\lambda_t = -W, \text{ for h=t} \qquad (10.A.4)$$

$$= 0, \text{ for h}\neq t \qquad (10.A.5)$$

which yields:

$$\partial B^{-1}/\partial \lambda_t = -E^{tt} \otimes W \tag{10.A.6}$$

$$[\partial B^{-1}/\partial \lambda_t].B = -E^{tt} \otimes W(I-\lambda_t W)^{-1} \tag{10.A.7}$$

with E^{tt} defined as before.

As a result, the LHS in the first order condition (10.A.1) reduces to the trace of the product of two block diagonal matrices, of which one has only one non−zero element. Consequently, the LHS simplifies to:

$$\text{tr } \{(\partial B^{-1}/\partial \lambda_t).B\} = \text{tr } \{-W(I-\lambda_t W)^{-1}\}.$$

Using the results (10.A.4) and (10.A.5), it follows that the elements of z_{ht} (10.A.3) take on a particular form, as in:

$$z_{ht} = -We_t, \text{ for } h=t$$

$$= 0, \text{ for } h \neq t$$

Consequently, the N by T matrix Z_t will consist of zero elements, except for column t, which equals $-We_t$. This greatly simplifies the RHS expression in (10.A.1):

$$\text{tr } Z\Sigma^{-1}Z_t' = \text{tr } [Z_t'Z].\Sigma^{-1}$$

The cross product is a T by T matrix of zero elements except for row t, which becomes $z_{tt}'z_h$, for h=1,...,T. Therefore, the trace can be obtained as the sum of the cross products of the elements of row t of $Z_t'Z$ with the elements of column t of Σ^{-1}, which yields:

$$\text{tr } Z\Sigma^{-1}Z_t' = \Sigma_h \sigma^{th}.[z_{tt}'.z_h)$$

$$= \Sigma_h \sigma^{th}.[e_t'W'(I-\lambda_h W)e_h]$$

as in (10.8).

Appendix 10.B.:   **Information Matrix for the ML Estimator in the Spatial SUR Model with Spatially Autocorrelated Errors.**

As in the Appendix 10.A, the information matrix for the ML estimator can be derived as a special case of the results in Magnus (1978). There, the information matrix for the general model with $\Omega(\theta)$ is of the form:

$$\Psi_{ij} = (1/2)\text{tr}[(\partial\Omega^{-1}/\partial\theta_i)\Omega(\partial\Omega^{-1}/\partial\theta_j)\Omega]$$

for all combinations of θ_i and θ_j in θ.

In the spatial SUR model, the information matrix is block diagonal between the elements of β and those of $[\lambda, \sigma]$. The result for β is the usual $X'\Omega^{-1}X$. The important elements of the information matrix for the parameters in Ω (λ and σ), say Ψ, are:

$$\Psi(\lambda,\lambda) = \text{tr } K_iK_j + \text{tr } K_i'(\Sigma^{-1} \otimes I)K_j(\Sigma \otimes I)$$

$$\Psi(\lambda,\sigma) = \text{tr } K_i'(\Sigma^{-1}E^{hk} \otimes I)$$

$$\Psi(\sigma,\sigma) = (1/2)N.\text{tr } (\Sigma^{-1}E^{ij})(\Sigma^{-1}E^{hk})$$

with i,j,h,k, as the relevant elements of the coefficient vectors, E defined as before, and K as an auxiliary matrix for each element λ_t of λ. In the spatial SUR model, K_t becomes:

$$K_t = (\partial B^{-1}/\partial\lambda).B$$

$$= -E^{tt} \otimes W(I-\lambda_t W)^{-1}$$

as in (10.A.7).

The expressions needed to derive $\Psi(\lambda,\lambda)$ and $\Psi(\lambda,\sigma)$ can be obtained by using the simplifying notation $D_t = W(I-\lambda_t W)^{-1}$ and the following intermediate results:

$$K_tK_s = (E^{tt} \otimes D_t)(E^{ss} \otimes D_s)$$

$$= (E^{tt}.E^{ss}) \otimes (D_tD_s)$$

or,

$$K_tK_s = 0, \text{ for } t\neq s$$

$$= E^{tt} \otimes (D_t)^2, \text{ for } t=s$$

$$\text{tr } K_tK_t = [\text{tr } E^{tt}].[\text{tr } (D_t)^2]$$

$$= \text{tr } (D_t)^2$$

$$K_t{}'(\Sigma^{-1} \otimes I)K_s(\Sigma \otimes I)$$

$$= (E^{tt} \otimes D_t{}')(\Sigma^{-1} \otimes I)(E^{ss} \otimes D_s)(\Sigma \otimes I)$$

$$= (E^{tt}\Sigma^{-1}E^{ss}\Sigma) \otimes (D_t{}'D_s)$$

$$\operatorname{tr}\,(E^{tt}\Sigma^{-1}E^{ss}\Sigma) = \sigma^{ts}.\sigma_{ss}$$

$$\operatorname{tr}\,K_t{}'(\Sigma^{-1} \otimes I)K_s(\Sigma \otimes I) = \sigma^{ts}.\sigma_{ts}.\operatorname{tr}\,(D_t{}'D_s)$$

$$K_t{}'(\Sigma^{-1}E^{hk} \otimes I) = (E^{tt} \otimes D_t{}')(\Sigma^{-1}E^{hk} \otimes I)$$

$$= (E^{tt}\Sigma^{-1}E^{hk}) \otimes D_t{}'$$

$$\operatorname{tr}\,K_t{}'(\Sigma^{-1}E^{hk} \otimes I) = [\operatorname{tr}\,(E^{tt}\Sigma^{-1}E^{hk})].[\operatorname{tr}\,D_t]$$

Appendix 10.C.:   **ML Estimator in the Two−Equation Spatial SUR Model with Spatially Autocorrelated Errors**

The estimation equations for the spatial SUR model with two equations are:

− for β:

$$\beta = [X'\Omega^{-1}X]^{-1}X'\Omega^{-1}y$$

or, in terms of a partitioning into the variables in each equation:

$$X'\Omega^{-1}X = \begin{bmatrix} \sigma^{11}X_1'(I-\lambda_1 W)'(I-\lambda_1 W)X_1 & \sigma^{12}X_1'(I-\lambda_1 W)'(I-\lambda_2 W)X_2 \\ \sigma^{21}X_2'(I-\lambda_2 W)'(I-\lambda_1 W)X_1 & \sigma^{22}X_2'(I-\lambda_2 W)'(I-\lambda_2 W)X_2 \end{bmatrix}$$

$$X'\Omega^{-1}y = \begin{bmatrix} \sigma^{11}X_1'(I-\lambda_1 W)'(I-\lambda_1 W)y_1 & \sigma^{12}X_1'(I-\lambda_1 W)'(I-\lambda_2 W)y_2 \\ \sigma^{21}X_2'(I-\lambda_2 W)'(I-\lambda_1 W)y_1 & \sigma^{22}X_2'(I-\lambda_2 W)'(I-\lambda_2 W)y_2 \end{bmatrix}$$

− for λ:

$$\text{tr } W(I-\lambda_1 W)^{-1} = \sigma^{11}e_1'W'(I-\lambda_1 W)e_1 + \sigma^{12}e_1'W'(I-\lambda_2 W)e_2$$

$$\text{tr } W(I-\lambda_2 W)^{-1} = \sigma^{21}e_2'W'(I-\lambda_1 W)e_1 + \sigma^{22}e_2'W'(I-\lambda_2 W)e_2$$

− for Σ:

$$\Sigma = (1/N) \begin{bmatrix} e_1'(I-\lambda_1 W)'(I-\lambda_1 W)e_1 & e_1'(I-\lambda_1 W)'(I-\lambda_2 W)e_2 \\ e_2'(I-\lambda_2 W)'(I-\lambda_1 W)e_1 & e_2'(I-\lambda_2 W)'(I-\lambda_2 W)e_2 \end{bmatrix}$$

The elements of the information matrix for β are as usual. For the elements of $[\lambda, \sigma]$ the corresponding expressions in the two equation case are:

$$\Psi(\lambda_1,\lambda_1) = \text{tr } (D_1)^2 + \sigma^{11}.\sigma_{11}.\text{tr } (D_1'D_1)$$

$$\Psi(\lambda_1,\lambda_2) = \sigma^{12}.\sigma_{12}.\text{tr } (D_1'D_2)$$

$$\Psi(\lambda_1,\sigma_{11}) = \sigma^{11}.\text{tr } (D_1)$$

$$\Psi(\lambda_1,\sigma_{12}) = \sigma^{12}.\text{tr } (D_1)$$

$$\Psi(\lambda_1,\sigma_{22}) = 0$$

$$\Psi(\lambda_2,\lambda_2) = \text{tr } (D_2)^2 + \sigma^{22}.\sigma_{22}.\text{tr } (D_2'D_2)$$

$$\Psi(\lambda_2,\sigma_{11}) = 0$$

$$\Psi(\lambda_2,\sigma_{12}) = \sigma^{21}.\text{tr } (D_2)$$

$$\Psi(\lambda_2,\sigma_{22}) = \sigma^{22}.\text{tr } (D_2)$$

$$\Psi(\sigma_{11}, \sigma_{11}) = (1/2)N.(\sigma^{11})^2$$

$$\Psi(\sigma_{11}, \sigma_{12}) = N.\sigma^{11}.\sigma^{12}$$

$$\Psi(\sigma_{11}, \sigma_{22}) = (1/2)N.(\sigma^{12})^2$$

$$\Psi(\sigma_{12}, \sigma_{12}) = N.[(\sigma^{12})^2 + \sigma^{11}.\sigma^{22}]$$

$$\Psi(\sigma_{12}, \sigma_{22}) = N.\sigma^{22}.\sigma^{12}$$

$$\Psi(\sigma_{22}, \sigma_{22}) = (1/2)N.(\sigma^{22})^2$$

Appendix 10.D.:   **The LM Test for Spatial Error Autocorrelation in the Two−Equation Spatial SUR Model.**

The LM test is of the form:

$$LM = \iota'(\Sigma^{-1}*U'WU)[T_1.I + T_2.\Sigma^{-1}*\Sigma]^{-1}(\Sigma^{-1}*U'WU)'\iota$$

or,

$$LM = \iota'PS^{-1}P'\iota$$

with

$$P = \Sigma^{-1}*U'WU$$

$$S = T_2.I + T_1.\Sigma^{-1}*R$$

and, as before,

$$T_1 = tr\ W'W$$

$$T_2 = tr\ W^2$$

For the two−equation case, the statistic is asymptotically distributed as $\chi^2(2)$. Its constituent parts can be written explicitly as:

$$P = \begin{bmatrix} \sigma^{11}e_1'We_1 & \sigma^{12}e_1'We_2 \\ \sigma^{21}e_2'We_1 & \sigma^{22}e_2'We_2 \end{bmatrix}$$

$$S = \begin{bmatrix} T_2 + T_1\sigma^{11}.\sigma_{11} & T_1\sigma^{12}.\sigma_{12} \\ T_1\sigma^{21}.\sigma_{21} & T_2 + T_1\sigma^{22}.\sigma_{22} \end{bmatrix}$$

Appendix 10.E.: **Derivation of the LM Test for Spatial Autocorrelation in a 2ECM.**

Under the assumption of normality, the log likelihood for the 2ECM with spatial autocorrelation is:

$$L = -(NT/2)\ln \sigma -(1/2) \ln |\Psi| -(1/2\sigma)e'\Psi^{-1}e$$

with

$$\Psi = \iota_T.\iota_T' \otimes \omega.I_N + I_T \otimes (B'B)^{-1}$$

$$\omega = \sigma(\mu)/\sigma$$

$$B = I - \lambda W$$

The score for λ is:

$$\partial L/\partial\lambda = -(1/2) \partial(\ln |\Psi|)/\partial\lambda -(1/2\sigma).e'(\partial\Psi^{-1}/\partial\lambda)e$$

In order to obtain the expression in (10.9), the following partial results are needed:

$$\partial(\ln|\Psi|)/\partial\lambda = \text{tr} \ \Psi^{-1}.(\partial\Psi/\partial\lambda)$$

$$\partial\Psi^{-1}/\partial\lambda = -\Psi^{-1}(\partial\Psi/\partial\lambda)\Psi^{-1}$$

$$\partial\Psi/\partial\lambda = I_T \otimes \partial(B'B)^{-1}/\partial\lambda$$

$$\partial(B'B)^{-1}/\partial\lambda = (B'B)^{-1}(W'B+B'W)(B'B)^{-1}$$

and, thus:

$$\partial(\ln|\Psi|)/\partial\lambda = -\text{tr} \ \Psi^{-1}.[I_T \otimes (B'B)^{-1}(W'B+B'W)(B'B)^{-1}]$$

$$\partial\Psi^{-1}/\partial\lambda = -\Psi^{-1}[I_T \otimes (B'B)^{-1}(W'B+B'W)(B'B)^{-1}]\Psi^{-1}$$

which lead directly to the complete expression for the score.

Under the null hypothesis, Ψ^{-1} takes on a simpler form, which allows for the terms in the score to be written more succintly. Indeed,

$$\Psi^{-1} = I_{NT} - E \otimes \kappa.I_N$$

$$\partial\Psi^{-1}/\partial\lambda = I_T \otimes W'+W$$

with

$$E = \iota_T.\iota_T'$$

$$\kappa = \omega/1+T\omega$$

Consequently, the first term in the score becomes:

$-(1/2) \; \partial(\ln |\Psi|)/\partial\lambda$

$= (1/2)\mathrm{tr} \; [I_{NT} - E \otimes \kappa.I_N].[I_T \otimes (W'+W)]$

$= (1/2)\mathrm{tr}[I_T \otimes (W+W')] - (1/2)\mathrm{tr}[E \otimes \kappa(W+W')]$

$= (1/2)\mathrm{tr}I_T.\mathrm{tr}(W+W') - (1/2)\mathrm{tr}E.\mathrm{tr}(W+W')$

$= 0$

since $\mathrm{tr}W = \mathrm{tr}W' = 0$ by construction (zero diagonal).

The second term in the score simplifies as well:

$(1/2\sigma)e'\Psi^{-1}(\partial\Psi/\partial\lambda)\Psi^{-1}e$

$= (1/2\sigma)e'[I_{NT} - E \otimes \kappa.I_N][I_T \otimes (W'+W)][I_{NT} - E \otimes \kappa.I_N]e$

Since this expression is scalar, and all matrices except W and W' are symmetric, it further reduces to:

$(1/\sigma)e'[I_{NT} - E \otimes \kappa.I_N][I_T \otimes W][I_{NT} - E \otimes \kappa.I_N]e$

$= (1/\sigma)e'\{[I_T \otimes W]-[E \otimes 2\kappa W]+[T.E \otimes \kappa^2 W]\}e$

$= (1/\sigma)e'\{[I_T + \kappa(T\kappa-2)E] \otimes W\}e$

using $E.E = T.E$.

The off-diagonal elements of the information matrix that pertain to λ and $[\sigma \; \omega]$ are:

$I(\lambda,\sigma) = (1/2\sigma).\mathrm{tr} \; \Psi^{-1}.(\partial\Psi/\partial\lambda)$

$I(\lambda,\omega) = (1/2)\mathrm{tr} \; \Psi^{-1}(\partial\Psi/\partial\lambda).\Psi^{-1}(\partial\Psi/\partial\omega)$

As shown above, $\mathrm{tr} \; \Psi^{-1}.(\partial\Psi/\partial\lambda)$ (i.e., the same as the first element of the score for λ) becomes zero under the null hypothesis. Also, since

$\partial\Psi/\partial\omega = E \otimes I_N$

and using the simplifying expressions for Ψ^{-1} and $\partial\Psi/\partial\lambda$ under the null, the second element becomes:

$I(\lambda,\omega) \quad = \quad (1/2)\mathrm{tr} \; [I_{NT} - E \otimes \kappa I_N][I_T \otimes (W+W')]$

$\qquad\qquad\qquad x[I_{NT} - E \otimes \kappa I_N][E \otimes I_N]$

$$= (1/2)\text{tr} \{[[I_T \otimes (W+W')][I_{NT} - E \otimes \kappa I_N][E \otimes I_N]$$

$$-(1/2)\text{tr}[E \otimes \kappa I_N][I_T \otimes (W+W')][I_{NT} - E \otimes \kappa I_N][E \otimes I_N]$$

The first term in this expression becomes:

$$(1/2)\text{tr} \{E \otimes (W+W') - T.E \otimes \kappa(W+W')\} = 0$$

and the second term becomes:

$$(1/2)\text{tr} \{T.E \otimes \kappa(W+W') - T^2.E \otimes \kappa^2(W+W')\} = 0$$

based on $\text{tr}(W+W')=0$.

Consequently, under the null hypothesis, the information matrix is block diagonal in the elements corresponding to λ.

NOTES ON CHAPTER 10

[1]   For ease of notation, in this chapter covariances and variances will be expressed as σ, instead of σ^2.

[2]   The Kronecker product $A \otimes B$ of the m by m matrix A and the n by n matrix B is an mn by mn matrix with as elements $a_{ij}.B$. In other words, each element in A is multiplied by the matrix B. Properties of the Kronecker product that will be used in the remainder of this section are: $(A \otimes B)(C \otimes D) = AC \otimes BD$; $(A \otimes B)^{-1} = A^{-1} \otimes B^{-1}$; $\det(A \otimes B) = [\det(A)]^n.[\det(B)]^m$; $\text{tr} (A \otimes B) = \text{tr}A.\text{tr}B$.

[3]   For details, see, e.g., Phillips (1985). See also the general comments in Section 8.2.1. for a discussion of the properties of EGLS estimators.

[4]   A specialized approach to the two–equation case is outlined in Kariya (1981).

[5]   For ease of exposition, the same weight matrix W is assumed for all time periods. The extension to the case of a different W_t for each t can be carried out in a straightforward way.

[6]   Note that here B equals the inverse of $(I-\lambda W)$, rather than $(I-\lambda W)$ itself, as in the previous chapters.

[7]   The formal derivation of this general case is given in Magnus (1978). See also Breusch (1980) for some important invariance results. The derivation of the ML etimator for the spatial SUR model was introduced in Anselin (1980, Chapter 7).

[8]   In order for the resulting estimates to be acceptable, the λ should satisfy stability requirements (e.g., $|\lambda_t|<1$) to avoid the indication of explosive spatial processes. In addition, the estimates for Σ should result in a positive definite matrix.

[9]   More specifically, the crucial transformation matrix $W(I-\lambda_t W)^{-1}$ fails to yield a zero trace in the spatial case, due to the structure of W. In time series, this matrix has zero diagonal elements, and thus yields a trace of zero. This ensures block diagonality between the information matrix partitions that correspond to the different parameters.

[10]   The resulting estimates for γ should satisfy the usual stability conditions, e.g., $|\gamma_t|<1$.

[11]   See also Anselin (1980), and Hooper and Hewings (1981) for some concerns about the interpretation of these space–time Moran coefficients.

[12]   The Hadamard or direct product of a matrix A and a matrix B (of the same dimensions) is obtained by multiplying each element of A with the corresponding element of B: $[A*B]_{ij} = a_{ij}.b_{ij}$.

[13]   Provided that each cluster has the same number of elements. If the clusters are of unequal size, the simplifying Kronecker product results of the SUR model no longer apply, although this situation can still be considered as a case of a nonspherical error variance. For a discussion of this issue in a time series context, see Schmidt (1977).

14   For a detailed discussion, see Hsiao (1986), as well as the references cited in the introduction to Section 10.2.

15   In a standard econometric approach, this dependence would be in the time domain, as in Lillard and Willis (1978).

16   Magnus (1982, p. 245) shows that for a matrix with special structure $U = i_T.i_T' \otimes C + I_T \otimes D$, the inverse and determinant can be written in a simpler form. With $M = (1/T)i_T.i_T'$, and $Q = D + T.C$, the matrix is equivalent to: $U = (M \otimes Q) + (I_T - M) \otimes D$. The determinant then follows as $|U| = |Q|.|D|^{T-1}$ and the inverse as $U^{-1} = M \otimes Q^{-1} + (I-M) \otimes D^{-1}$.

17   For example, see the extensive discussion and derivations in Hsiao (1986).

18   For notational simplicity, only spatial equations are considered here. The simultaneous framework can easily be extended to include other non–spatial equations (and variables as well).

19   The special structure considered by Magnus (1978) is of the form $\Omega = Q(\Sigma \otimes \Delta)Q'$.

CHAPTER 11

PROBLEM AREAS IN ESTIMATION AND TESTING FOR

SPATIAL PROCESS MODELS

In this chapter I will briefly review some fundamental methodological problems associated with estimation and testing in spatial process models. The attention paid to these issues in the spatial literature varies. Some problems, such as those resulting from pre–testing, have essentially been ignored. Others, such as the boundary value issue, have resulted in a considerable body of literature, but have not yet been resolved in a satisfactory manner. In general, the issues discussed in this chapter can be considered to form important directions for future research, crucial to an effective operational implementation of spatial econometric techniques.

The chapter consists of four sections. In the first, the pre–test problem associated with the interpretation of spatial tests and estimators is considered. The second section is devoted to the boundary value problem. In the third section, some brief remarks are formulated on the issue of the specification of the spatial weight matrix. The chapter is concluded with a discussion of problems related to the sample size in the operational implementation of spatial estimators.

11.1. Pre–Testing

In many contexts in applied econometrics, the choice of an estimator is conditional upon the result of a previously carried out specification test or regression diagnostic. For example, in spatial analysis, a rejection of the null hypothesis of no spatial autocorrelation in the error term would typically lead to a maximum likelihood estimation that takes the spatial dependence into account.

This sequential approach changes the stochastic properties of the estimator and has important consequences for hypothesis testing and prediction. The issue is known as the pre–test problem in econometrics, and has received considerable attention, e.g., in Wallace (1977), Judge and Bock (1978) and Judge and Yancey (1986). The particular case of pre–test estimation in the presence of serial correlation in the regression errors has been analyzed in, e.g., Nakamura and Nakamura (1978), Griffiths and Beesley (1984), King and Giles (1984), and Giles and Beattie (1987). In spatial analysis, this issue has so far essentially been ignored.

In the remainder of this section, I will first outline in general terms the formal approach underlying pre–test estimators. This is followed by a brief discussion of pre–test estimators in spatial process models.

11.1.1. General Framework

The selection of an estimator in function of the outcome of a hypothesis test is approached within the framework of statistical decision theory. In general terms, the problem is considered as a decision about the value of the parameters of a probability distribution, when the true value is unknown. This decision is based on the observation of random data, generated by the underlying unknown distribution. The estimator itself is the formal decision rule which determines a value for the parameter in function of the data.

The appropriateness of a particular decision rule or estimator is judged by means of an objective function, or loss function, which formulates the quantifiable consequences of the choice of the estimate in function of the underlying true value. Typically, this is expressed in terms of a measure of distance between the true value and the estimate, such as an absolute error or a squared error. For example, a weighted squared error loss function would be:

$$L[\theta, \theta(y)] = [\theta(y) - \theta]'A[\theta(y) - \theta]$$

where θ is a vector of parameters, $\theta(y)$ its estimate, expressed as a function of the random data y, and A is a general weight matrix. Since the data y are random, the loss will also be random. Therefore it is often measured by its expected value, or risk:

$$R[\theta, \theta(y)] = E\{L[\theta, \theta(y)]\}$$

where the expected value is over the distribution $f(y|\theta)$, i.e., the density of the data conditional upon the true parameter vector.

A pre−test estimator is a decision rule which consists of a stochastic mixture of alternative estimators. The random nature of the mixture is caused by the uncertainty associated with a hypothesis test, the outcome of which determines the choice of the estimator.

A common example is the case of a nonspherical error covariance matrix which is a function of a parameter, say θ. If θ is zero, OLS is the best linear unbiased estimator. If θ is nonzero, estimation should be based on a consistent estimator, EGLS or ML. The decision about whether or not θ is zero depends on the interpretation of the probability level associated with a hypothesis test. In other words, if the test statistic is considered to be significant, EGLS is used, otherwise OLS is appropriate. However, even if the test indicates significance, there is a probability α (the type I error) that its value is generated by the null hypothesis and thus that the wrong estimator is used. The consequences of this decision are measured by the risk function, which in general depends on the true parameter values, the data, the significance level of the test, and the choice of the test statistic.

Formally, the choice between the OLS and EGLS estimators b_{OLS} and b_{EGLS} depends on the outcome of a test on the following null hypothesis:

$$H_0: \theta = 0$$

with as alternative $H_1: \theta \neq 0$, or possibly, $H_1: \theta > 0$.

The decision is based on the value for a test statistic, $t(y)$. When $t(y)$ exceeds a pre−set critical value c, for a given significance level α, the null hypothesis is rejected. Formally, with α and c such that:

$$P[t(y) \geq c] = \alpha$$

the following decision rule results:

$t(y) \geq c$	→ reject H_0		→ use b_{EGLS}
$t(y) < c$	→ do not reject H_0		→ use b_{OLS}

The resulting pre−test estimator b_{PTE} can be expressed in function of the outcome of the test $t(y)$ and critical level $c(\alpha)$ as:

$$b_{PTE} = I_R(\alpha).b_{EGLS} + I_{NR}(\alpha).b_{OLS}$$

where $I_R(\alpha)$ and $I_{NR}(\alpha)$ are indicator functions, such that:

$$I_R(\alpha) = 1 \text{ for } t(y) \geq c(\alpha)$$

$$I_R(\alpha) = 0 \text{ otherwise}$$

and

$$I_{NR}(\alpha) = 1 \text{ for } t(y) < c(\alpha)$$

$$I_{NR}(\alpha) = 0 \text{ otherwise.}$$

A strategy of always using OLS, i.e., ignoring the possibility that the nuisance parameter θ may be non−zero, corresponds to a choice of $\alpha=0$. In other words, the null hypothesis is never rejected. The opposite strategy of always using EGLS, even when θ may be zero, corresponds to a choice of $\alpha=1$, i.e., the null hypothesis is always rejected.

A variety of pre−test estimators can be developed by choosing a different test statistic, a different significance level, or a different alternative estimator. Their distributional properties are complex and very few analytical results have been obtained. Most findings are based on Monte Carlo simulation. They indicate that the properties of pre−test estimators depend on the data matrix, the particular test statistic and the pre−set significance level. [1]

The pre−test approach illustrates how the usual inference based on the distributional properties of b_{OLS} or b_{EGLS} will be misleading when the choice of this estimator is based on the outcome of a hypothesis test. Different estimation strategies can be compared in terms of risk, for varying magnitudes for the nuisance parameter θ and significance levels α. [2] It turns out that under certain circumstances, the pre−test estimator is dominated in terms of risk by a class of biased Stein−rule or shrinkage estimators. A further discussion of this issue is beyond the current scope. [3]

11.1.2. Pre—Test Estimators in Spatial Process Models

As illustrated in Chapters 6 and 8, estimation becomes considerably more complex when the error terms in a regression are spatially dependent, or when a spatially lagged dependent term is included. The necessity of this more complex estimation is often first checked by means of a test for the presence of spatial effects. In this way, the estimation of spatial process models becomes embedded in a pre—test framework.

This pre—test aspect of estimation in spatial process models has been virtually ignored in the literature, even though its dynamic counterpart has been extensively discussed. Much research remains to be done to gain insight into the relative risks associated with alternative spatial estimation strategies and to assess the potential for other approaches such as Stein—rule estimators. Of particular applied interest in this respect would be an assessment of the extent to which OLS in the presence of spatial dependence is dominated by other estimators in terms of risk, for a relevant range of autocorrelation coefficients and spatial dependence structures.

A number of issues are specific to the spatial case and cannot be directly transposed from the results in a time series context. As outlined in Section 8.1, the tests for spatial error dependence are different from those used for serial autocorrelation. Also, the significance levels associated with the Moran, Wald, Likelihood Ratio and Lagrange Multiplier tests are not exact, but based on asymptotic considerations. On the other hand, only two acceptable EGLS procedures are available, i.e., the full ML and the Spatial Durbin approach, in contrast to the wide range of alternatives in time series.

11.2. The Boundary Value Problem

An important methodological problem that affects statistical inference in spatial process models results from the fact that spatial dependence may transcend the boundaries of the data set. In other words, if the value of a variable measured for a spatial unit depends on its values in other spatial units in the data set, it may also depend on values in spatial units that are not included. Typically, this problem is referred to as the boundary value problem or edge effect, although it is not limited to the border in a strict sense.

This problem is similar to the starting value issue in time series analysis, but it differs in a number of important respects. In space, the number of edges is much larger than the single origin in time. Moreover, the designation of spatial units as boundary units is not always unambiguous in irregular spatial structures. Also, the dependence in space is multidirectional rather than one—directional. This implies that the values outside the sample not only influence those in the sample (i.e., the usual edge effect), but are also influenced in their turn by those in the sample, yielding a complex simultaneous pattern of dependence.

The issue of edge effects has received substantial attention in the spatial literature, and has been addressed in the context of spatial process modeling by Haining (1978d), Haggett (1980, 1981), and more recently in the work of Griffith (Griffith and Amrhein 1983, and Griffith 1983, 1985). In addition, it constitutes an important problem in point pattern analysis, as discussed in Ripley (1981) and Diggle (1983). However, for the type of data structures relevant in spatial econometrics, the discussion of edge effects in the literature is fraught with definitional problems and has not yet resulted in satisfactory solutions, as illustrated by the recent interchange between Martin (1987) and Griffith (1988a).

Next, I will briefly outline in general terms the main issues and correction techniques that have been suggested in the literature. This is followed by a more formal approach to edge effects, based on a spatial econometric viewpoint.

11.2.1. The Problem and its Correction

The concern with edge effects in spatial analysis is mostly focused on the bias that may result in the estimation of measures of spatial dependence. Typically, the discussion is in the limited context of the estimation of a spatial autocorrelation measure or of the parameter in a spatial autoregressive process, for a regular square lattice structure. [4]

Although the property of unbiasedness is important, its relevance in this context can be questioned, since it may not be attained by the usual maximum likelihood estimation process, even in the absence of edge effects. Indeed, as pointed out in Section 6.2, the ML approach is based on asymptotic considerations. When the usual regularity conditions are satisfied, it will yield consistent or asymptotically unbiased estimates. However, in many situations ML estimators will not achieve this unbiasedness in finite samples.

The border effects are formalized in terms of a first order contiguity between the spatial units along the four edges of a square lattice, and corresponding spatial units immediately outside of the square lattice. Although the values of the pertinent variable in the surrounding units are assumed to be unobserved, the location of the bordering units and the underlying pattern of dependence (weight matrix) are considered to be known. As illustrated by Griffith and Amrhein (1983), the existence of this type of edge effect results in a biased estimate of spatial dependence when ignored.

Several solutions or corrections for the border effect have been suggested. They are classified by Griffith (1983) into traditional and contemporary or statistical solutions. For example, one of the traditional approaches consists of a transformation of the lattice structure into an infinite torus, thereby creating an artificial contiguity between opposite edges in the square. Also, a variety of approaches based on the creation of buffer zones have been suggested, where the observations in the buffer zone are dropped from the analysis, but their dependence with the data in the sample is taken into account. [5] As shown in several simulation experiments in Griffith and Amrhein (1983), none of the traditional approaches provides a satisfactory correction mechanism.

Three statistical approaches are suggested in Griffith (1983) and analyzed in simulation experiments in Griffith (1985). Each correction technique approaches the boundary value problem from a different perspective. One incorporates it into a generalized least squares framework and introduces corrections similar to those in the case of serial error autocorrelation. Another focuses on the inhomogeneity introduced by the edge effects, which is modeled by a dummy variable for the boundary units. The third technique is based on a spatial interpolation perspective and reconstructs values for a buffer zone based on a maximum likelihood approach to missing values. The simulation results failed to show that any of these approaches resulted in satisfactory adjustments. [6]

11.2.2. A Spatial Econometric Approach to Edge Effects

In order to gain additional insight into the complexities of the boundary value problem in the estimation of spatial process models, it is useful to consider a more general framework than first order contiguity on a regular square lattice. In general, edge effects occur when spatial dependence is present between variables or error terms of the observed spatial units, and those in unobserved spatial units. In this context, observed spatial units are considered as a subset of a potentially infinitely large spatial system in which spatial dependence is present.

Formally, a random variate y is observed in spatial units $i,j \in I$, and unobserved in spatial units $h,k \in H$, where $I \cup H$ makes up the total spatial system S, and $I \cap H = \varnothing$. Spatial covariance is present when $E[y_i.y_j] \neq 0 \; \forall \; i,j$ for which $d_{ij} < \delta$, where d_{ij} is the distance between i and j according to a well defined metric. [7] If the same spatial process holds for the complete spatial system S, then also $E[y_h.y_k] \neq 0 \; \forall \; h,k$ for which $d_{hk} < \delta$. Furthermore, edge effects are generated for those units $i \in I$ and $h \in H$, for which $d_{ih} < \delta$, and $E[y_i.y_h] \neq 0$.

From a spatial econometric perspective, two different aspects of the consequences of edge effects can be distinguished. One pertains to the situation where spatial dependence is present in the error term only, the other is relevant when the regression model contains a spatially lagged dependent variable.

The first case can be illustrated without loss of generality for a spatial autoregressive process in the error:

$$\epsilon = \lambda W \epsilon + \mu$$

where ϵ is a vector of error terms partitioned according to I and H (e.g., with N spatial units in I and M units in H), W is an overall spatial weight matrix and μ is an independent error vector. In partitioned form, this becomes:

$$\begin{bmatrix} \epsilon_I \\ \epsilon_H \end{bmatrix} = \lambda \begin{bmatrix} W_{II} & W_{IH} \\ W_{HI} & W_{HH} \end{bmatrix} \begin{bmatrix} \epsilon_I \\ \epsilon_H \end{bmatrix} + \begin{bmatrix} \mu_I \\ \mu_H \end{bmatrix}$$

where the partitioned elements of W indicate the underlying dependence within I and H, and for spatial units across the boundaries between I and H. The usual transformation yields the elements of ϵ in function of the independent variates μ, as:

$$\begin{bmatrix} \epsilon_I \\ \epsilon_H \end{bmatrix} = \begin{bmatrix} I - \lambda W_{II} & -\lambda W_{IH} \\ -\lambda W_{HI} & I - \lambda W_{HH} \end{bmatrix}^{-1} \begin{bmatrix} \epsilon_I \\ \epsilon_H \end{bmatrix} + \begin{bmatrix} \mu_I \\ \mu_H \end{bmatrix}$$

or, after partioned inversion, with

$$B_I = I-\lambda W_{II}$$

$$B_H = I-\lambda W_{HH}$$

the elements of ϵ_I follow as:

$$\epsilon_I = [B_I-\lambda^2 W_{IH}(B_H)^{-1}W_{HI}]^{-1}\mu_I$$

$$+\lambda[B_I-\lambda^2 W_{IH}(B_H)^{-1}W_{HI}]^{-1}W_{IH}(B_H)^{-1}\mu_H \qquad (11.1)$$

Expression (11.1) illustrates two important aspects of edge effects. As expected, the values for μ_H are introduced into the expression for ϵ_I. However, in addition, and in contrast to the situation in time series, the interaction between I and H also alters the form of the spatial dependence between the spatial units in I, as shown by the first term in the expression. [8] This is due to the two—directional nature of spatial dependence, and is often ignored in the discussion of spatial edge effects. Also, (11.1) indicates how all elements from μ_H exert an influence on the elements in ϵ_I, and not just the immediate border units. Of course, in order to satisfy the necessary regularity conditions for ML, this influence must decrease with increasing distance.

When border effects are taken into account, the error covariance for ϵ_I becomes considerably more complex than the usual $\sigma^2(B_I'B_I)^{-1}$. Based on (11.1), assuming independence of μ, and with:

$$D = [B_I-\lambda^2 W_{IH}(B_H)^{-1}W_{HI}]$$

the error covariance $\Omega=E[\epsilon_I.\epsilon_I']$ is found as:

$$\Omega = \sigma^2(D'D)^{-1} + \sigma^2[\lambda^2(D)^{-1}W_{IH}(B_H'B_H)^{-1}W_{IH}'(D')^{-1}]$$

Clearly, the standard approach of ignoring edge effects results in a severely misspecified error covariance matrix.

Although the above expressions illustrate the complexities introduced by edge effects, they are not operational, since W_{HH}, W_{IH} and W_{HI} are unknown. It is obvious that ignoring this additional complexity is not likely to yield consistent estimates for λ. The extent of this problem in realistic contexts is not known. Further research is needed to take into account this source of misspecification by developing consistent approximations or robust procedures, e.g., as discussed in Section 8.3.

When a spatially lagged dependent variable is included in the regression, edge effects can be considered as a special form of functional misspecification. Indeed, the estimated relation contains only part of the spatial dependence. Formally, with y_I and y_H partitioned as before:

$$y = \rho W_{II}y_I + \rho W_{IH}y_H + X\beta + \mu$$

or

$$y = \rho W_{II} y_I + X\beta + v$$

$$v = \mu + \rho W_{III} y_H$$

where the unobserved spatial dependence is included in the error term v.

Unlike the μ, the new error term v is not likely to have a zero mean and is no longer spherical. [9] Furthermore, due to the dependence between y_I and y_H, the error v will not be uncorrelated with the lagged dependent variable. Again, since W_{III} is unknown, it is not possible to obtain an operational maximum likelihood estimator for the parameters in this expression. The extent of the effect of this misspecification on the consistency and other properties of the usual estimators (ML or IV) is not known. In addition, further research is needed to obtain consistent estimators that may be robust to the misspecification. One promising approach is the general optimization of a quasi—likelihood as suggested by White (1982a), although further insight is needed to properly incorporate the two—directional nature of dependence in space.

Alternatively, to the extent that edge effects are more pronounced near the boundaries of the spatial data set, the influence of the missing spatial dependence could be incorporated as a special case of heterogeneity. Indeed, procedures based on varying coefficients or structural change may be able to address some of the misspecification by classifying spatial units into distinct sub—zones, as outlined in Chapter 9.

Also, since the error term v has nonzero mean, an indication of the importance of the edge effects could be obtained from the extension to spatial process models of general specification tests similar to the RESET approach of Ramsey (1974).

11.3. The Specification of the Spatial Weight Matrix

As pointed out in Chapter 3, the spatial weight matrix is a formal expression of constraints on the full two—dimensional pattern of spatial dependence. The imposition of these constraints is necessitated by the lack of information in cross sectional data to estimate a complete set of $N(N+1)$ spatial interaction coefficients. When observations are available over time as well as across space, these constraints can be relaxed. In the particular case where the time dimension is larger than the spatial dimension, a spatial weight matrix is no longer necessary, although it may still be desired for particular spatial processes, such as Markov processes. In addition, the use of a weight matrix may result in more efficient estimates of the parameters.

An important methodological problem in the estimation of spatial process models is associated with the proper choice of the weights. A misspecified weight matrix may result in inconsistent estimates and misleading inference, as illustrated in Stetzer (1982b). Moreover, the uncertainty associated with the value of the weights, which is typically ignored, may affect the interpretation of the estimation results.

A framework to properly incorporate the inherent stochastic nature of the weights is important, since competing weight matrices can often be considered for the same empirical context, with no clear a priori criteria to guide the choice. A Bayesian approach would seem to be most appropriate, although to date very little progress has been made in this direction.

In sum, inference in spatial process models is very susceptible to misspecifications in the weight matrix. This can only be avoided by the use of robust approaches. These are still in the early stages of development and the full implications of their application are not well understood. Consequently, it is crucial to be able to assess the validity of the assumptions embedded in the weight matrix. This is further explored in Chapters 13 and 14.

11.4. The Importance of Sample Size for Spatial Process Estimators

The number of observations affects the operational estimation of spatial process models in two important respects. One issue is associated with small sample sizes, the other with a large number of observations.

Since the properties of estimators and tests for spatial process models are based on asymptotic considerations, their application in small samples may be problematic. As pointed out in the previous chapters, significance levels derived from asymptotic normality may be misleading. In addition, the asymptotic equivalence of the Wald, Likelihood Ratio and Lagrange Multiplier approaches fails to be reflected in finite samples and may lead to conflicting indications. Also, the normality of the distribution of the estimated parameters may not carry over in small samples.

Much remains to be done in order to obtain better insight into the importance of these issues. One possible avenue for research would consist of the further development of finite sample approximations. Alternatively, inference could be based on Bayesian strategies or on resampling procedures such as the bootstrap, as in Section 7.3. In addition, a much larger body of simulation results is needed to provide better guidance in the choice of estimators and tests for spatial process models in realistic contexts.

In very large data sets operational problems of a different nature occur. This is due to the fact that estimators and tests based on the maximum likelihood approach to spatial process models contain weight matrices and Jacobians of a dimension equal to the number of observations. Consequently, numerical problems associated with the inversion and computation of determinants for very large (and sparse) matrices will limit the size of the data sets that can be considered. These limitations are likely to slowly disappear as computational technology advances and supercomputing becomes more accessible. However, at this point in time, they effectively preclude some large data sets from being analyzed from a spatial econometric perspective.

In a number of instances the use of a large spatial weight matrix can be avoided. For example, as illustrated in Chapter 10, the presence of data in both time and space dimensions may allow for the spatial dependence to be estimated directly or to be subsumed in a general covariance matrix of unspecified form.

Also, when the spatial structure satisfies sufficient regularity conditions with respect to stationarity and isotropy, a spatial time series approach or a spectral analysis perspective can be taken. Finally, for some types of problems, the spatial dependence can be expressed in a hierarchical or nested form, as suggested in Section 10.1.5. Here too, substantial additional research is needed before more operational approaches become available for use in applied empirical regional science.

NOTES ON CHAPTER 11

[1]   Instead of using analytical or simulation results, inference can also be based on Bayesian notions or on resampling strategies, such as the bootstrap.

[2]   Typically, these comparisons are based on Monte Carlo simulations, e.g., in King and Giles (1984), Griffiths and Beesley (1984) and Giles and Beattie (1987).

[3]   For a detailed discussion, see in particular Judge and Bock (1978) and Judge and Yancey (1986).

[4]   Although, as Martin (1987) has noted, the underlying general spatial process framework within which the edge effects are analyzed is not always made explicit.

[5]   This is somewhat similar to the approach in time series analysis, where the first observation in a first order autoregressive process is dropped, but taken into account as the lagged dependent variable for the second observation.

[6]   Based on the simulation experiments, Griffith suggested that an approach based on the generalized least squares framework would be most promising. For a more detailed discussion, see Griffith (1983, 1985, 1988a) and also Martin (1987).

[7]   See also Chapter 3.

[8]   Without edge effects the transformation would be $\epsilon_I = (B_I)^{-1} \mu_I$.

[9]   A zero mean for ν would only obtain in the special case where $E[y_H] = 0$, or in the trivial case where $\rho = 0$.

CHAPTER 12

OPERATIONAL ISSUES AND EMPIRICAL APPLICATIONS

To round off the discussion in Part II, in this chapter several of the estimation methods and tests for spatial process models that were previously developed in formal terms will be illustrated empirically. Three aspects are considered in particular. In the first section, I focus on some operational issues related to the implementation of maximum likelihood estimation and the associated nonlinear optimization problem. In the second section, the analysis of cross—sectional data is considered. Using a simple model of determinants of crime for 49 contiguous neighborhoods in Columbus, Ohio, several estimation methods and tests from Chapters 6, 8 and 9 are implemented empirically. In the third section, attention shifts to space—time data sets. A Phillips curve model is estimated at two points in time for 25 contiguous counties in South—Western Ohio, to illustrate the instrumental variable methods from Chapter 7 and various diagnostics for spatial effects from Chapter 10.

12.1. Operational Issues

The implementation of the maximum likelihood estimators for spatial process models necessitates a nonlinear optimization of the likelihood function in terms of several parameters. Although this is a familiar problem in nonlinear estimation in econometrics, some particular characteristics of spatial models merit special attention. In this section, some of these aspects are considered in more detail.

First, the precise form of the optimization problem needed for the implementation of ML estimators is illustrated for two special cases of the spatial process model discussed in Chapter 6. It turns out that the maximization of the likelihood function can be reduced to an expression in one parameter, conditional upon the values of the other parameters, which simplifies matters greatly. Next, the numerical problems associated with the presence of a Jacobian term in the likelihood are briefly discussed. In the third and fourth part of this section, the practical relevance is evaluated of direct search and other numerical approaches to the optimization in spatial models. The section is concluded with a brief assessment of the availability of appropriate software.

12.1.1. Maximum Likelihood Estimation

For the general spatial process model outlined in Chapter 6, the ML estimates are obtained from a maximization of the following log—likelihood function with respect to the parameters β (K coefficients), α (P+1 coefficients), ρ and λ:

$$L = -(N/2)\ln\pi - (1/2)\ln|\Omega| + \ln|A| + \ln|B|$$

$$-(1/2)(Ay-X\beta)'B'\Omega^{-1}B(Ay-X\beta)$$

with, as before,

$$A = I-\rho W$$

$$B = I-\lambda W.$$

Two special cases of this model are of particular interest in applied work: the mixed regressive spatial autoregressive model (i.e., with $B=I$ and $\Omega=\sigma^2 I$), and the linear regression model with spatially dependent errors (i.e., with $A=I$ and $\Omega=\sigma^2 I$). In both cases, the maximization of the log–likelihood reduces to a simpler expression in the form of a concentrated likelihood in the spatial parameter, conditional upon the values for the other coefficients.

For the mixed regressive spatial autoregressive model, the log–likelihood becomes:

$$L = -(N/2)\ln\pi - (N/2)\ln\sigma^2 + \ln|A| - (1/2\sigma^2)(Ay-X\beta)'(Ay-X\beta)$$

The application of the general first order conditions from Chapter 6 (expressions 6.21–6.24) to this special case yields b as the estimator for β, for which:

$$b = (X'X)^{-1}X'Ay$$

or

$$b = (X'X)^{-1}X'y - \rho(X'X)^{-1}X'Wy = b_0 - \rho.b_L$$

The OLS estimators b_0 and b_L are obtained from a regression of the X on y and on Wy respectively. Clearly, the ML estimate for β is a function of these auxiliary regression coefficients as well as of ρ. However, whereas the estimate for ρ cannot be expressed analytically, neither b_0 nor b_L are a function of any other parameters. Therefore, the estimate for β can be found directly, once a value for ρ has been determined.

The two coefficient vectors b_0 and b_L lead to two sets of residuals, e_0 and e_L, which depend on the X and y (and Wy) only:

$$e_0 = y - Xb_0$$

$$e_L = Wy - Xb_L.$$

Further application of the first order conditions, and taking into account the auxiliary residuals yields the estimate for the error variance σ^2 as:

$$\sigma^2 = (1/N)(e_0-\rho e_L)'(e_0-\rho e_L)$$

Again, this estimate can be readily obtained once a value for ρ has been determined.

Substitution of the estimates for β and σ^2 into the likelihood results in a concentrated likelihood of the following form:

$$L_C = C - (N/2)\ln[(1/N)(e_O-\rho e_L)'(e_O-\rho e_L)] + \ln|I-\rho W|$$

where C is the usual constant. This expression is a nonlinear function in one parameter only, namely ρ, and can be easily maximized by means of the numerical techniques that will be outlined in Sections 12.1.3. and 12.1.4.

Consequently, the estimation process can proceed according to the following steps:

[1]   carry out OLS of X on y: yields b_O

[2]   carry out OLS of X on Wy: yields b_L

[3]   compute residuals e_O and e_L

[4]   given e_O and e_L, find ρ that maximizes L_C

[5]   given ρ, compute $b=b_O-\rho b_L$ and $\sigma^2=(1/N)(e_O-\rho e_L)'(e_O-\rho e_L)$.

All except step [4] in this approach can be carried out by means of a standard regression package. Step [4] necessitates the use of an appropriate nonlinear optimization routine.

Estimation is slightly more complex in the regression model with spatially dependent error terms. The corresponding log−likelihood is of the form:

$$L = -(N/2)\ln\pi -(N/2)\ln\sigma^2 + \ln|B| -(1/2\sigma^2)(y-X\beta)'B'B(y-X\beta)$$

and the familiar first order conditions lead to the EGLS estimate for β:

$$b_{EGLS} = [X'B'BX]^{-1}X'B'By$$

and the estimate for σ^2 as:

$$\sigma^2 = (1/N)e'B'Be$$

with $e=y-Xb_{EGLS}$.

The estimators for β and σ^2 are both a function of the value of λ. Using the same approach as above, a concentrated log−likelihood can be obtained as a nonlinear function in this parameter:

$$L_C = C - (N/2)\ln[(1/N)e'B'Be] + \ln|I-\lambda W|$$

However, in contrast to the situation for the mixed regressive spatial autoregressive model, the residual vector e in L_C is indirectly also a function of λ, since b_{EGLS} is obtained for a value of λ. Therefore, a one−time optimization of L_C with respect to λ does not suffice to obtain ML estimates for all parameters. Indeed, a numerically complex simultaneous approach is needed, or a less demanding iterative procedure can be followed. This would essentially alternate back and forth between the estimation of λ conditional upon a vector of residuals e (generated for a value of b_{EGLS}), and an estimation of β (and $σ^2$) conditional upon a value for λ, until convergence is obtained.

Estimation should then proceed according to the following steps:

[1]   carry out OLS of X on y: yields b_{OLS}

[2]   compute initial set of residuals $e=y-Xb_{OLS}$

[3]   given e, find λ that maximizes L_C

[4]   given λ, carry out EGLS: yields b_{EGLS}

[5]   compute new set of residuals $e=y-Xb_{EGLS}$

[6]   if convergence criterion is met, proceed to [7], else return to [3]

[7]   given e and λ, compute $σ^2=(1/N)e'B'Be$.

Although these steps are slightly more complex than in the previous case, all but step [3] in this procedure can be carried out by means of a standard regression package, by using OLS for a set of transformed variables $X^*=(I-λW)X$ and $y^*=(I-λW)y$. As before, step [3] necessitates a nonlinear optimization routine. [1]

Essentially the same iterative approach can be taken when the error variance structure is expressed in function of several parameters. [2] For example, this would apply to the model with joint spatial dependence and heteroskedasticity and in the SUR approach with spatial dependence in the errors. In each case, the optimization of a concentrated likelihood in terms of the variance parameters (α in the heteroskedastic model, Σ in SUR) can be carried out conditional upon the value for the spatial parameter. In practice, this results in the use of a spatially adjusted residual vector $u=(I-λW)e$ and the application of a standard approach to obtain the variance terms. For example, in the spatial SUR model, estimation would proceed along the following steps, with the index t referring to each equation:

[1]   estimate each equation by OLS: yields b_{OLS}

[2]   compute initial set of equation−specific residuals, e_t

[3]   given e_t, find $λ_t$ in each equation which optimizes the corresponding concentrated likelihood (i.e., the standard approach for a spatially dependent error term)

[4]   given the $λ_t$, construct the spatially adjusted residual vectors $u_t=(I-λ_t W)e_t$

[5]   derive an estimate for Σ based on the u_t (expression 10.4. in Chapter 10)

[6]   given λ_t and Σ, obtain the spatial SUR estimate for β by applying a standard SUR estimation on the spatially transformed variables $X^* = (I - \lambda_t W)X$ and $y^* = (I - \lambda_t W)y$

[7]   check convergence criterion: if met, stop, else go to [8]

[8]   obtain a new set of residuals $e_t = y_t - b_{t(SUR)}$

[9]   given e_t and Σ, derive a new estimate for the λ_t from the joint first order conditions (expression 10.8 in Chapter 10)

[10]   given λ_t, compute a new set of adjusted residuals u_t

[11]   given u_t, derive Σ

[12]   go to [6]

Steps 3 and 9 in this iterative procedure require specialized optimization routines. However, the other steps can be carried out by using properly transformed variables in a standard econometric regression package.

12.1.2. The Jacobian

The likelihood function for the general spatial process model, as well as the concentrated likelihood for the special cases discussed above, contains one or two Jacobian terms, $\ln|A|$ and/or $\ln|B|$. Since the matrices A and B are of dimension equal to the number of observations, their presence in the function to be optimized considerably complicates the numerical analysis. Indeed, the determinant and its derivative (which involves a matrix inversion) need to be evaluated at each iteration for a new value of the spatial parameter ρ or λ.

In practice, two different approaches can be taken. In one, following an early suggestion by Ord (1975), the determinant is expressed in function of the eigenvalues of the spatial weight matrix, such that:

$$\ln|I - \rho W| = \ln \Pi_i (1 - \rho.\omega_i) = \Sigma_i \ln (1 - \rho.\omega_i)$$

where the ω_i are the eigenvalues of W. The first partial derivative of this expression with respect to the spatial parameter ρ becomes:

$$\partial\ln|I - \rho W|/\partial\rho = \partial[\Sigma_i \ln (1 - \rho.\omega_i)]/\partial\rho = \Sigma_i -\omega_i/1 - \rho\omega_i$$

Consequently, at each iteration, the determinant and its partial derivative can be computed as a straightforward sum of N terms. Since the eigenvalues are not affected by the value of ρ, they need only be calculated once, which results in a gain in computing speed and a reduction in computer memory requirements. However, except for some special regular lattice structures, the eigenvalues of the weight matrix cannot be obtained analytically and need to be derived by means of a numerical procedure. More specifically, a routine is needed which computes the eigenvalues of a general real but potentially non−symmetric matrix. Although several such routines are available in commercial statistical and mathematical packages, the precision of the results decreases rapidly as the size of the matrix increases. In other words, especially for larger data sets, the gain in speed and conservation of memory may be partially offset by a loss in precision. [3]

An alternative approach consists of a brute force calculation of the determinant and the inverse matrix at each iteration. [4] For most reasonably sized problems, the use of routines which are optimized for these matrix manipulations will result in higher precision and a negligible loss in speed. [5] However, when the number of observations becomes very large, the numerical requirements for these operations can only be reliably satisfied in a supercomputing environment.

12.1.3. The Direct Search Approach to Optimisation

The concentrated likelihood for the spatial process model is a nonlinear function of one parameter only. It therefore can be optimized in a straightforward fashion by direct search methods. This is facilitated even more by the requirement that stable values for the spatial parameter are restricted to the range -1 to $+1$. Consequently, a local maximum can be obtained by evaluating the concentrated likelihood (or its derivative) for a series of values of the spatial coefficient at small intervals over the range -1 to $+1$. Subsequent searches around this initial optimum, at increasingly finer intervals will eventually lead to an estimate with the desired level of precision.

For a spatial autoregressive process, this procedure can be speeded up considerably by taking advantage of the known properties of the likelihood, and particularly its behavior for values of ρ close to one. In Appendix 12.A, a simple algorithm is outlined for a bisection search that implements this approach. Specifically, the choice of the initial upper and lower limit for the parameters is based on the relative magnitude of the derivative of the likelihood for the OLS estimate and the Moran coefficient. Subsequent values are derived as the midpoint between the upper and lower limit, in the usual fashion.

This simple algorithm yields a local optimum with the desired level of precision in a small number of steps. It is particularly suited for estimation in the mixed regressive spatial autoregressive model, where one series of iterations suffices. For models with spatial dependence in the error term, this algorithm is less appropriate, since it does not exploit the convergence properties of the other parameters (i.e., the b_{EGLS}) at successive iterations. However, it can still be applied to obtain a reasonable starting value for use in another optimization method.

12.1.4. Other Optimisation Methods

In models with spatial dependence in the error term, a direct search approach is not efficient, since it needs to be carried out for every iteration of b_{EGLS} estimates. Other, more traditional nonlinear optimization techniques, such as a steepest decent method, a Gauss Newton approach, or a Davidon Fletcher Powell procedure can be applied in a straightforward manner, using the first and/or second derivatives for the respective concentrated likelihoods.

To different extents, these methods are sensitive to the choice of the initial value for the parameters. For poor choices of these values, they may fail to converge or lead to unacceptable estimates, such as values for ρ larger than one. The Gauss Newton approach in particular tends to lead to such results and has not been found to be very useful for this particular type of problem.

For the models with spatial dependence in the error that are illustrated in the remainder of this chapter, estimation is based on an application of the steepest descent method. In this approach, the parameter value at each iteration is adjusted in function of the magnitude of the derivative of the concentrated likelihood. However, to keep the parameter change from yielding values outside the acceptable range (-1 to $+1$), the derivative is adjusted by a scaling coefficient. Formally, a new value at iteration i is obtained as:

$$\rho_i = \rho_{i-1} + s.d(\rho_{i-1})$$

where $d(\rho_{i-1})$ is the partial derivative of the concentrated likelihood and s is a scale factor. In the examples that follow, a scale factor that limits parameter adjustments to 0.01 of the derivative ensures non−explosive behavior for the algorithm. [6]

12.1.5. Software Availability

To a large extent, the dissemination of spatial econometric techniques in the practice of regional science has been hampered by a lack of readily available software, since none of the more familiar statistical and econometric computer packages contain a designated set of procedures for spatial analysis. [7] As pointed out above, many aspects of the estimation of spatial econometric models can be carried out by means of standard methods on properly transformed data sets. However, the implementation of the nonlinear optimization and the iteration back and forth between the estimation of the spatial coefficient and the other parameters requires special programming. Sometimes this can be incorporated into a commercial econometric package, by using macro facilities and efficient procedures for specialized matrix manipulations, such as matrix inversion, the computation of the determinant, and the extraction of eigenvalues. [8] Typically, special programming efforts have been carried out for specific studies, which can be found in the form of technical reports and academic discussion papers, with only limited dissemination. [9]

The estimation and testing of the various models presented in the following two sections was carried out by means of an integrated set of routines written in the GAUSS matrix programming language for IBM compatible microcomputers. A full listing of the respective procedures could not be included here because of space limitations, but can be found in detail in Anselin (1987c). All computations were carried out in a microcomputer environment, on a Compaq Portable (8086) and an IBM AT (80286) equipped with a mathematical coprocessor.

12.2. The Analysis of Cross—Sectional Data

In this section, various ways of dealing with spatial dependence and spatial heterogeneity in cross—sectional data sets are illustrated empirically. After a brief introduction of the model and data, a total of eight different situations are considered, four dealing with spatial dependence (based on Chapters 6 and 8), and four dealing with combined dependence and heterogeneity (based on Chapters 8 and 9). Spatial dependence is illustrated in diagnostic tests for OLS estimation, in ML estimation of a mixed regressive spatial autoregressive model, ML estimation of a model with spatially dependent error terms, and in the spatial Durbin approach. Spatial heterogeneity is shown in an application of the spatial expansion method, in tests for spatial dependence in a heteroskedastic model, and tests for structural stability in a spatially dependent model. The treatment of cross—sectional data is concluded with an illustration of ML estimation of the model with an error term that incorporates joint spatial dependence and heteroskedasticity.

12.2.1. A Spatial Model of Determinants of Neighborhood Crime

The model that will be used throughout in the treatment of cross—sectional data is a simple linear expression relating crime to measures of income and housing value. This model is selected primarily to illustrate the various spatial effects, and is not intended to contribute to a substantive understanding of spatial patterns of crime. The model is estimated using observations for 49 contiguous Planning Neighborhoods in Columbus, Ohio. These neighborhoods correspond to census tracts, or aggregates of a small number of census tracts, and are representative of the type of data used in many empirical urban analyses. Their spatial layout is illustrated in Figure 12.1.

The crime variable (CRIME) pertains to the combined total of residential burglaries and vehicle thefts per thousand households in the neighborhood. Income (INC) and housing values (HOUSE) are in thousand dollars. All data pertain to 1980. A complete listing is presented in Table 12.1.

Spatial effects are incorporated in two ways: spatial dependence as expressed by a weight matrix, and spatial heterogeneity in the form of heteroskedasticity, spatial parameter variation and spatial structural shifts. Dependence is taken as first order contiguity between the neighborhoods, as illustrated in Table 12.2. A row—standardized weight matrix is used in all analyses. Heteroskedasticity is expressed as a linear functional relation between the error variance and the squares of the explanatory variables. It thus corresponds to a special form of random coefficient variation. Spatial parameter variation is formulated in the expansion method as a function of the neighborhood centroid coordinates. These are listed as variables X and Y in Table 12.1. A spatial structural shift is taken into account by distinguishing between neighborhoods east and west of a main north—south transportation axis, expressed by means of a dummy variable, listed as EAST in Table 12.1.

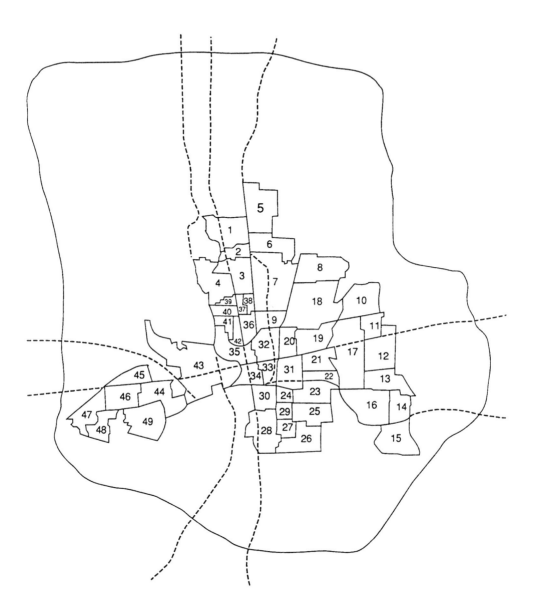

Figure 12.1. Neighborhoods in Columbus, Ohio

Table 12.1. Determinants of Crime, Columbus Ohio Neighborhoods

NEIG	CRIME	INCOME	HOUSING	X	Y	EAST
1	18.802	21.232	44.567	35.62	42.38	0
2	32.388	4.477	33.200	36.50	40.52	0
3	38.426	11.337	37.125	36.71	38.71	0
4	0.178	8.438	75.000	33.36	38.41	0
5	15.726	19.531	80.467	38.80	44.07	1
6	30.627	15.956	26.350	39.82	41.18	1
7	50.732	11.252	23.225	40.01	38.00	1
8	26.067	16.029	28.750	43.75	39.28	1
9	48.585	9.873	18.000	39.61	34.91	1
10	34.001	13.598	96.400	47.61	36.42	1
11	36.869	9.798	41.750	48.58	34.46	1
12	20.049	21.155	47.733	49.61	32.65	1
13	19.146	18.942	40.300	50.11	29.91	1
14	18.905	22.207	42.100	51.24	27.80	1
15	27.823	18.950	42.500	50.89	25.24	1
16	16.241	29.833	61.950	48.44	27.93	1
17	0.224	31.070	81.267	46.73	31.91	1
18	30.516	17.586	52.600	43.44	35.92	1
19	33.705	11.709	30.450	43.37	33.46	1
20	40.970	8.085	20.300	41.13	33.14	1
21	52.794	10.822	34.100	43.95	31.61	1
22	41.968	9.918	23.600	44.10	30.40	1
23	39.175	12.814	27.000	43.70	29.18	1
24	53.711	11.107	22.700	41.04	28.78	1
25	25.962	16.961	33.500	43.23	27.31	1
26	22.541	18.796	35.800	42.67	24.96	1
27	26.645	11.813	26.800	41.21	25.90	1
28	29.028	14.135	27.733	39.32	25.85	1
29	36.664	13.380	25.700	41.09	27.49	1
30	42.445	17.017	43.300	38.32	28.82	1
31	56.920	7.856	22.850	41.31	30.90	1
32	61.299	8.461	17.900	39.36	32.88	1
33	60.750	8.681	32.500	39.72	30.64	1
34	68.892	13.906	22.500	38.29	30.35	0
35	38.298	14.236	53.200	36.60	32.09	0
36	54.839	7.625	18.800	37.60	34.08	0
37	56.706	10.048	19.900	37.13	36.12	0
38	62.275	7.467	19.700	37.85	36.30	0
39	46.716	9.549	41.700	35.95	36.40	0
40	57.066	9.963	42.900	35.72	35.60	0
41	54.522	11.618	30.600	35.76	34.66	0
42	43.962	13.185	60.000	36.15	33.92	0
43	40.074	10.655	19.975	34.08	30.42	0
44	23.974	14.948	28.450	30.32	28.26	0
45	17.677	16.940	31.800	27.94	29.85	0
46	14.306	18.739	36.300	27.27	28.21	0
47	19.101	18.477	39.600	24.25	26.69	0
48	16.531	18.324	76.100	25.47	25.71	0
49	16.492	25.873	44.333	29.02	26.58	0

Table 12.2. First Order Contiguity, Columbus, Ohio Neighborhoods.

NEIGHBORHOOD	CONTIGUOUS TO:								
1:	2	5	6						
2:	1	3	6	7					
3:	2	4	7	37	38	39			
4:	3	37	39	40					
5:	1	6							
6:	1	2	5	7					
7:	2	3	6	8	9	18	36	38	
8:	7	18							
9:	7	18	20	32	36	38			
10:	11	17	18	19					
11:	10	12	17						
12:	11	13	17						
13:	12	14	16	17					
14:	13	15	16						
15:	14	16							
16:	13	14	15	17	23				
17:	10	11	12	13	16	18	19	21	22	23
18:	7	8	9	10	17	19	20		
19:	10	17	18	20	21	31			
20:	9	18	19	31	32				
21:	17	19	22	31					
22:	17	21	23	31					
23:	16	17	22	24	25	29	31		
24:	23	25	29	30	31	33			
25:	23	24	26	27	29				
26:	25	27	28						
27:	25	26	28	29					
28:	26	27	29	30					
29:	23	24	25	27	28	30			
30:	24	28	29	31	33	34			
31:	19	20	21	22	23	24	30	32	33
32:	9	20	31	33	34	35	36		
33:	24	30	31	32	34	35			
34:	30	32	33	35	43				
35:	32	33	34	36	41	42	43		
36:	7	9	32	35	37	38	40	42	
37:	3	4	36	38	39	40	42		
38:	3	7	9	36	37				
39:	3	4	37	40					
40:	4	36	37	39	41	42			
41:	35	40	42						
42:	35	36	37	40	41				
43:	34	35	44						
44:	43	45	46	49					
45:	44	46							
46:	44	45	47	48	49				
47:	46	48							
48:	46	47							
49:	44	46							

12.2.2. OLS Regression with Diagnostics for Spatial Effects

The application of ordinary least squares to the regression of crime on a constant term, income and housing values yields the following results, with the estimated standard deviation and corresponding t−values listed in parentheses:

$$CRIME = \begin{array}{ccc} 68.619 & -1.597 \text{ INC} & -0.274 \text{ HOUSE} \\ (4.735) & (0.334) & (0.103) \\ (14.490) & (4.781) & (2.654) \end{array}$$

$$R^2 = 0.552 \qquad\qquad \sigma^2 = 130.759$$
$$R_a^2 = 0.533 \qquad\qquad \sigma^2_{ML} = 122.753$$

All estimated coefficients are strongly significant. Moreover, the regression achieves a reasonable fit in terms of an R^2 and adjusted R^2 (R_a^2) of more than 0.5. Two estimates for the residual variance are listed, one unbiased, σ^2, and one maximum likelihood, σ^2_{ML}. Since the latter is obtained by dividing the residual sum of squares by the total number of observations, instead of the degrees of freedom (for the unbiased estimate), it will necessarily be smaller.

A traditional test for the presence of spatial autocorrelation in the error term is based on the Moran statistic computed for the residuals, as outlined in Section 8.1.1. For the regression above, this statistic is $I=0.236$. Its interpretation depends on whether adjustments are made for the particular nature of residuals, or whether a simple randomization approach is taken. In the former case, the relevant estimate for the mean and standard deviation are -0.033 and 0.091 respectively. In the latter case, these estimates are -0.021 and 0.093. Consequently, the standardized Moran statistics turn out to be $z_I=2.954$ for the normal case and $z_I=2.765$ for the randomization case, both highly significant with a probability of more than 0.99.

A full set of diagnostics for both spatial dependence and heteroskedasticity can be based on the Lagrange Multiplier approach, as outlined in Section 6.3.4. The results for the various one−directional and multidirectional tests are listed in Table 12.3. as χ^2, with q as the corresponding degrees of freedom and p as the associated probability level. Clearly, there is strong evidence of multiple sources of misspecification. The highest significance is achieved for a test against an omitted spatial lag, although both error dependence and error heterogeneity are indicated as well.

The LM test for spatial autocorrelation in the error term has a slightly lower significance than that indicated by a traditional Moran test, although this is not a general result. Also, as pointed out in Section 8.1.1, there is a formal relation between the two tests, in that the LM value can be found as $(N.I)^2/T$, where $T=trace[W^2 + W'W]$. For the weight matrix corresponding to the data in Table 12.2. this trace equals 23.294. Consequently, the LM statistic is $(49 \times 0.236)^2/23.294 = 5.723$.

Table 12.3. Lagrange Multiplier Diagnostics for Spatial Effects in
OLS Regression

	q	χ^2	p
One Directional Tests			
Spatial Error Autocorrelation	1	5.723	0.02
Omitted Spatial Lag	1	9.364	0.002
Random Coefficient Variation	2	7.900	0.02
Multidirectional Tests			
Spatial Dependence	2	9.443	0.01
Error Autocorrelation			
and Heteroskedasticity	3	13.624	0.003
All Effects	4	17.344	0.002

12.2.3. ML Estimation of a Mixed Regressive Spatial Autoregressive Model

The estimation results for a mixed regressive spatial autoregressive model, which includes a spatially lagged variable W_CRIME are reported in Table 12.4. The values obtained in an initial ordinary least squares regression are listed, as well as ML estimates, asymptotic standard deviations and an asymptotic t−test (i.e., the square root of the Wald statistic) for the significance of each variable. The estimated asymptotic variance matrix for all coefficients, including the error variance σ^2, is given as well. This matrix can be used to carry out asymptotic significance tests based on the Wald statistic, for various linear combinations of the parameters.[10]

The ML estimate for the spatial autoregressive coefficient is obtained from a simple bisection search, illustrated in Table 12.5. For each value of ρ, the associated value of the partial derivative of the concentrated log−likelihood is listed as well. The starting value of 0.3344 is the midpoint between the OLS estimate 0.5573 and 0.1115 (i.e., OLS times 0.2, as explained in Appendix 12.A). Since the corresponding derivative is positive, and the derivative for the OLS estimate was negative, the next value is the midpoint between 0.3344 and 0.5573, namely 0.4459. The iteration proceeds until the difference between two subsequent estimates is less than the convergence criterion.[11]

The ML estimate for ρ, 0.431, is smaller in absolute value than the result obtained by OLS, 0.557. This is in general accordance with the results from Monte Carlo simulations.[12] The coefficients for the other variables in the regression are similar to the OLS results without a spatial lag, though smaller in absolute value. The determinant of $A=I-\rho W$, which gives an indication of the extent to which the spatial dependence affects the likelihood, is evaluated as 0.312.

All estimated coefficients are clearly significant according to the asymptotic t−test, or, equivalently, a Wald test. For the spatial autoregressive parameter, the latter is the square of the t−value, i.e., 13.415. A Likelihood Ratio test for ρ can

Table 12.4. Estimates in the Mixed Regressive Spatial Autoregressive Model

VARIABLE	OLS	ML	ST.DEV.	T
W_CRIME	0.557	0.431	0.118	3.663
CONSTANT	38.181	45.079	7.177	6.261
INC	−0.866	−1.032	0.305	3.381
HOUSE	−0.264	−0.266	0.089	3.005

L = −165.408
σ^2 = 95.495

Asymptotic Variance Matrix

W_CRIME	0.0139	−0.6976	0.0127	0.0009	−0.3236
CONSTANT	−0.6976	51.5143	−1.3260	−0.1616	16.3015
INC	0.0127	−1.3260	0.0931	−0.0118	−0.2959
HOUSE	0.0009	−0.1616	−0.0118	0.0078	−0.0203
σ^2	−0.3236	16.3015	−0.2959	−0.0203	379.7751

Table 12.5. Bisection Search for the Spatial Autoregressive Parameter

ITERATION	ρ	DERIVATIVE
1	0.3344	5.952558
2	0.4459	−0.979915
3	0.3901	2.610210
4	0.4180	0.845598
5	0.4319	−0.059602
6	0.4250	0.394894
7	0.4285	0.168119
8	0.4302	0.054377
9	0.4311	−0.002583
10	0.4306	0.025904
11	0.4308	0.011663
12	0.4310	0.004540
13	0.4310	0.000979

be based on the difference between L in Table 12.4. and the log–likelihood for the OLS estimation (without a spatially lagged variable) from the previous section. Since the OLS regression has an estimated residual variance of 122.753 (using the ML estimate), the corresponding log–likelihood (including the constant) follows as -170.395. Consequently, a Likelihood Ratio test for ρ yields a value of 9.974, which is also clearly significant at a probability level of over 0.99. The relative ranking of the magnitudes for the LM test, LR test and Wald statistic is in accordance with the theoretically expected result:

$$9.364 < 9.974 < 13.415$$

or, LM < LR < W, as in Section 6.3.5.

A final aspect to consider is the extent to which spatial error autocorrelation remains present after the introduction of the spatially lagged dependent variable. A traditional approach, although without a rigorous foundation, is based on the use of the Moran statistic computed for the residuals. In the example, this statistic is $I=0.038$, and its corresponding standardized z value (using a randomization assumption) is 0.651, clearly not significant.

As pointed out in Section 8.1.3, a more rigorous asymptotic approach can be derived from the Lagrange Multiplier principle. The resulting statistic can be computed from the Moran value, the estimated variance for ρ, and some auxiliary traces, as $(N.I)^2/T$, where $T=t_{22}-(t_{21A})^2 \cdot \text{var}(\rho)$. The first trace, $t_{22}=\text{tr}(W^2+W'W)$, and equals 23.294, as in the previous section,. The second trace is slightly more complex, $t_{21A}=\text{tr}(W+W')(WA^{-1})$, and can be found to be 29.990. From Table 12.4. the variance for ρ follows as 0.0139. Consequently, the LM statistic can be obtained as $(49\text{x}0.038)^2/[23.294-(29.990)^2\text{x}0.0139]=0.320$, which should be compared to the critical values in a χ^2 distribution with one degree of freedom. [13]

Clearly, no evidence is present of a spatial autoregressive error process after the introduction of a spatially lagged dependent variable.

12.2.4. Spatial Dependence in the Error Term

Since the LM tests in Table 12.3. also gave a strong indication of potential spatial dependence in the error term, this specification should be considered as well. The results of a maximum likelihood estimation are listed in Table 12.6. The EGLS estimate for the regression coefficients are given, as well as the estimated asymptotic standard deviation and asymptotic t–value. The estimate for λ is the value that maximizes the concentrated likelihood and achieves the desired convergence criterion.

The variance matrix for the estimates (including σ^2) is listed as well. The block–diagonal structure between the regression coefficients and the parameters of the error variance is obvious. [14]

In the first two iterations, the optimization of the concentrated likelihood, conditional upon the values for the residuals, is carried out by means of a simple bisection search. The starting values are taken as the OLS estimate for λ and the Moran coefficient, both computed for the residuals. In the initial iteration, these values are 0.655 and 0.236 respectively. Further iterations are carried out by means of a steepest descent method. The resulting estimates for λ that are used in successive EGLS procedures are listed in Table 12.7. together with the corresponding σ^2, which provides an indication of the improving overall fit. [15]

An asymptotic t−test or Wald test clearly indicates the significance of λ: the corresponding values are 4.197 (standarized normal variate) and 17.611 (χ^2 with 1 degree of freedom). As in the previous section, a Likelihood Ratio test for λ can be based on the likelihood in Table 12.6, L=−166.398, and the likelihood in the OLS regression. The resulting statistic is 7.994, which is also highly significant. As before, the ranking of the Wald, LR and LM test values is in accordance with theory:

$$5.723 < 7.994 < 17.611$$

or, LM<LR<W.

A remaining concern is whether the inclusion of a spatial autoregressive process in the error term eliminates the indication of heteroskedasticity. This can be assessed by means of the spatially adjusted Breusch−Pagan test outlined in Section 9.2.1. Its value is 19.520, which is highly significant for a χ^2 variate with 2 degrees of freedom. A Breusch−Pagan test that ignores spatial effects yields a value of 19.914, which is larger than the spatial version, as expected, but only marginally so, and would not result in a qualitatively different judgement. At any rate, there is a strong indication that both spatial dependence and heteroskedasticity may be present in the error term.

Table 12.6.   ML Estimation of the Model with Spatially Dependent Error Terms

VARIABLE	ML	ST.DEV.	T
CONSTANT	59.893	5.366	11.161
INC	−0.941	0.331	−2.848
HOUSE	−0.302	0.090	−3.341
λ	0.562	0.134	4.197

L   =   −166.398
σ^2 =   95.575

Asymptotic Variance Matrix

CONSTANT	28.7957	−0.9988	−0.1244	0.0000	0.0000
INC	−0.9988	0.1093	−0.0136	0.0000	0.0000
HOUSE	−0.1244	−0.0136	0.0082	0.0000	0.0000
λ	0.0000	0.0000	0.0000	0.0179	−0.6297
σ^2	0.0000	0.0000	0.0000	−0.6297	394.9640

Table 12.7.   Iterations in the ML Estimation

ITERATION	λ	σ^2
1	0.393	102.381
2	0.517	97.235
3	0.551	95.966
4	0.559	95.666
5	0.561	95.596
6	0.562	95.580
7	0.562	95.576

12.2.5. A Spatial Durbin Model

As pointed out in Section 8.2.2, an alternative approach to estimation in a model with a spatial error autoregression consists of the spatial Durbin method. This is equivalent to the application of the ML technique for a mixed regressive autoregressive model on a specification which includes the spatially lagged explanatory variables. The resulting estimates are presented in Table 12.8, where the initial OLS result is listed, as well as the final ML value, estimated variance and asymptotic t–value.

Whereas the estimated ρ parameter is clearly significant, the coefficients of the lagged explanatory variables are not. Moreover, for W_INC, the wrong sign is obtained, since the common factor hypothesis would imply a positive sign, given a positive estimate for ρ and a negative sign for INC. This provides some evidence that an omitted spatial lag may be the main spatial effect, rather than spatial dependence in the error term. [16]

A Lagrange Multiplier test on spatial error autocorrelation yields a nonsignificant value of 0.289 (χ^2 with one degree of freedom). A similar indication is provided by the Moran statistic, which has a standardized z–value of 0.382 (for $I=0.014$). Therefore, the inclusion of higher order spatial lags in the specification would not seem warranted.

12.2.6. Diagnostics for Spatial Effects in the Spatial Expansion Method

One of the ways in which spatial heterogeneity can be incorporated into the model is by means of a simple spatial expansion, discussed in Section 9.3. Four new explanatory variables are included, which are constructed by multiplying the original INC and HOUSE with the x and y coordinate of the neighborhood centroid. This linear expansion is only one of many possible specifications. Since the model is used here primarily for illustration purposes, a more complex formulation is not pursued.

Table 12.8. Estimation of the Spatial Durbin Model

VARIABLE	OLS	ML	ST.DEV.	T
W_CRIME	0.758	0.426	0.156	2.729
CONSTANT	17.780	47.822	12.667	3.381
INC	−0.763	−0.914	0.331	2.761
HOUSE	−0.297	−0.294	0.089	3.293
W_INC	0.141	−0.520	0.565	0.921
W_HOUSE	0.288	0.246	0.179	1.373

$$L = -164.411$$
$$\sigma^2 = 91.791$$

The results of an OLS estimation of the spatially expanded model are presented in Table 12.9. A number of alternative estimates for the coefficient standard deviation are listed as well, since, as pointed out in Section 9.3.2, heteroskedasticity is likely to be present due to potential error in the construction of the expanded variables. Specifically, three heteroskedasticity−robust measures from Section 8.3.3. are applied: the White estimate, a small sample adjusted White estimate, and a jackknife estimate. The influence of these estimates on the resulting inference is assessed in Table 12.10, where the probabilities are listed that are associated with the corresponding (asymptotic) t−tests on the significance of the coefficients.

The introduction of four new explanatory variables affects the fit of the model only to a marginal extent: the adjusted R^2 increases from 0.533 to 0.581. However, the coefficient estimates (in terms of value and sign) and their significance are substantially affected. Using the standard t−test approach, the expanded variables associated with the y coordinate (i.e., a west−east trend) are strongly significant, while those associated with the x coordinate (a south−north trend) are not. The coefficient for the non−expanded income variable has a much larger value than before, but is only marginally significant (p=0.057). The non−expanded coefficient for housing values changes sign and is no longer significant.

The interpretation of the significance of the model parameters changes drastically when the heteroskedasticity−robust measures are used. In general, the White variance leads to an indication of higher significance, while the other two measures point to less significance. In the jackknife approach, only the coefficient for the income variable remains marginally significant, which illustrates the potentially overly optimistic inference that the less robust methods may provide.

A final issue to assess is the extent to which the spatial expansion eliminates other spatial effects. The relevant Lagrange Multiplier statistics are listed in Table 12.11. Clearly, there is still a strong indication of the presence of spatial dependence, both in the form of error autocorrelation and in the form of an omitted spatially lagged dependent variable. A Moran statistic of 0.213 with associated z−

function of INC and HOUSE) seems no longer present, as measured by a Breusch–Pagan test. Consequently, the robust inference may not really be needed, although it may be capturing some of the spatial misspecifications. Nevertheless, since the inference is so strongly affected, a general caution should be exerted in the standard interpretation of the results.

Table 12.9. OLS Estimation of the Spatially Expanded Model

VARIABLE	COEFF	ST.DEV	WHITE	ADJ.WHITE	JACK
CONSTANT	69.505	4.599	4.017	4.476	5.106
INC	−4.091	2.092	1.593	1.880	2.279
HOUSE	0.405	0.779	0.520	0.648	0.845
X_INC	−0.046	0.034	0.030	0.040	0.052
X_HOUSE	0.027	0.013	0.012	0.016	0.022
Y_INC	0.121	0.055	0.053	0.068	0.089
Y−HOUSE	−0.049	0.021	0.021	0.029	0.039

R^2 = 0.633 $\qquad$ σ^2 = 117.305
R_a^2 = 0.581 $\qquad$ σ^2_{ML} = 100.547

Table 12.10. Heteroskedastic–Robust Inference in the Spatially Expanded Model

VARIABLE	P(T)	P(WHITE)	P(ADJ.WHITE)	P(JACK)
INC	0.06	0.01	0.04	0.08
HOUSE	0.61	0.44	0.54	0.63
X_INC	0.18	0.14	0.25	0.38
X_HOUSE	0.04	0.03	0.10	0.23
Y_INC	0.03	0.03	0.08	0.18
Y_HOUSE	0.02	0.03	0.10	0.22

Table 12.11. Lagrange Multiplier Diagnostics for Spatial Effects in
the Spatially Expanded Model

	q	χ^2	p
Spatial Error Autocorrelation	1	4.660	0.03
Omitted Spatial Lag	1	13.346	0.0003
Heteroskedasticity	2	0.743	0.69

12.2.7. Testing for Spatial Error Dependence in a Heteroskedastic Model

In a more tradtional fashion, heterogeneity can be incorporated in the form of a random coefficient model, discussed in Section 9.4.1. The resulting specification is a special case of a linear regression with a heteroskedastic error variance.

Two sets of results are reported. One set pertains to a model with the variables INC and HOUSE as the determinants of heteroskedasticity. This is the form used in the Breusch—Pagan test reported in Table 12.3. The other set includes only the variable HOUSE in the heteroskedastic variance, since the estimated random variance for INC turns out to be negative. The maximum likelihood estimates are presented in Table 12.12. [17]

The random variance component for HOUSE is only marginally significant, in that a one—sided asymptotic t—test (since the variance should be positive) achieves a probability level of 0.11. The component for INC is highly significant. However, since it has a negative sign, it is incompatible with the random coefficient model.

A test for spatial autocorrelation in the error term of this heteroskedastic model can be based on the Lagrange Multiplier approach outlined in Section 8.1.4. The resulting statistics are 8.822 and 6.084 for the two—component and one—component model respectively. Both are highly significant (for a χ^2 variate with one degree of freedom), indicating a persistent presence of spatial dependence. [18]

12.2.8. Testing for Structural Stability in the Presence of Spatial Error Dependence

An alternative perspective on spatial heterogeneity is to incorporate structural instability. In the model considered here, this is implemented by considering a separate regression for the east and west side of the city, as expressed by the dummy variable EAST. Some indication of such a spatial pattern of heterogeneity was also given by the significance of the expanded variables along the Y—coordinate in Section 12.2.6.

Table 12.12. ML Estimates for Random Coefficient Models

VARIABLE	COEFF	ST.DEV.	T	COEFF	ST.DEV.	T
CONSTANT	66.526	4.133	16.096	70.292	4.590	15.314
INC	−1.550	0.295	5.261	−1.718	0.344	4.991
HOUSE	−0.234	0.115	2.043	−0.274	0.117	2.354

Random Variance Components

VARIABLE	COEFF	ST.DEV.	T	COEFF	ST.DEV.	T
CONSTANT	129.478	24.506	5.284	93.663	23.547	3.978
INC	−0.173	0.073	2.377			
HOUSE	0.017	0.013	1.252	0.016	0.013	1.206

$$L = -168.384 \qquad\qquad L = -169.765$$

The estimation of the so−called unrestricted model, where a different set of coefficients is allowed for each subregion, is carried out by ML for a specification with a spatially dependent error. The estimates are listed in Table 12.13, which also includes the OLS results for the unrestricted model, needed for a traditional Chow test. The estimates for the so−called restricted model were given in Section 12.2.2. for OLS, and in Table 12.6. for ML. In those specifications the equality of coefficients over the subareas is imposed.

The ML results show a pattern quite distinct from the restricted estimates. Before, both INC and HOUSE had negative and significant coefficients. Now the significance varies by location: INC is only significant for the east, and HOUSE is only significant in the west. The insignificant coefficients also have a positive sign. The spatially autoregressive coefficient λ is larger than in the restricted model (compare 0.698 to 0.562) and is highly significant.

A formal test for the presence of structural instability can be carried out by means of a Chow statistic, based on the OLS estimation, and ignoring the spatial dependence in the error. The resulting statistic has a value of 1.823, which fails to be significant. [19]

Asymptotic spatially adjusted tests for structural instability can be based on the Wald, LR and LM statistics outlined in Section 9.2.2. The corresponding values are: LM=9.519, LR=12.227, W=15.398, which are all significant for a χ^2 variate with three degrees of freedom. Consequently, when spatial dependence is acknowledged, evidence is found for structural instability, whereas the traditional approach does not provide this indication.

Table 12.13. Estimation of the Unrestricted Model with Spatial Error Dependence

	COEFF	OLS	ML	ST.DEV.	T
EAST					
	CONSTANT	67.294	58.941	6.015	9.799
	INC	−2.014	−1.680	0.388	4.334
	HOUSE	−0.064	0.010	0.112	0.088
WEST					
	CONSTANT	76.650	55.904	7.755	7.209
	INC	−1.455	0.094	0.448	0.211
	HOUSE	−0.545	−0.576	0.112	5.139
λ			0.698	0.104	6.709

$$L = -160.285$$
$$\sigma^2 = 69.957$$

12.2.9. Joint Spatial Dependence and Heteroskedasticity in the Error Term

The final model considered for cross−sectional data is the specification with both spatial dependence and heteroskedasticity in the error term. Except for the omission of a spatially lagged dependent variable, this model is the general case discussed in Chapter 6. Since the variable INC resulted in a negative estimate for its random variance, only the variable HOUSE is included in the heteroskedastic term.

Estimation is carried out by maximum likelihood. The actual iterations consist of a combination of EGLS for the β coefficients, a steepest descent algorithm for the spatial autoregressive parameter, and the Amemiya (1985, p. 206) estimator for the random components, applied to the spatially transformed residuals. The detailed results for each iteration are listed in Table 12.14. The complete set of final parameter estimates are presented in Table 12.15. [20]

The estimated coefficients for both INC and HOUSE are negative and highly significant. The spatially autoregressive parameter is highly significant as well, but the random variance component associated with HOUSE is clearly not significant. A Likelihood Ratio test on the joint error parameters yields a value of 11.409, which rejects the null hypothesis of homoskedastic independent errors with a probability of over 0.99.

Table 12.14. Iterations for the Error Variance Parameters in the ML Estimation of the Model with Joint Spatial Dependence and Heteroskedasticity

ITERATION	λ	CONSTANT	HOUSE
1	0.376	62.639	0.025
2	0.583	68.610	0.012
3	0.629	67.665	0.010
4	0.643	67.378	0.010
5	0.647	67.304	0.010
6	0.648	67.286	0.010
7	0.648	67.281	0.010
8	0.648	67.280	0.010
9	0.648	67.279	0.010
10	0.648	67.279	0.010

Table 12.15. ML Estimation in the Model with Joint Spatial Dependence and Heteroskedasticity

VARIABLE	COEFF.	ST.DEV.	T
CONSTANT	58.677	5.349	10.970
INC	−0.816	0.335	2.435
HOUSE	−0.318	0.098	3.247

Parameters of the Error Variance

VARIABLE	COEFF.	ST.DEV.	T
λ	0.648	0.115	5.635
CONSTANT	67.279	23.530	2.859
HOUSE	0.010	0.013	0.781

$L = -164.691$

12.3. The Analysis of Space—Time Data

Several techniques outlined in Chapter 10, which address spatial effects in the analysis of space—time data are empirically illustrated in this section. Four aspects in particular are considered more closely: the estimation of a system of mixed regressive spatial autoregressive equations by means of instrumental variables methods, testing for the presence of spatial dependence in seemingly unrelated regressions and in the error components model, and the estimation of a SUR model with spatially autocorrelated error terms. This is preceded by a brief discussion of the data and model that will be used throughout, and an initial assessment of spatial effects, based on OLS estimation.

12.3.1. A Spatial Phillips—Curve

The model that will be used throughout the empirical application is a simple Phillips—curve specification, which relates changes in wage rates (WAGE) to the inverse unemployment rate (UN), and a net—migration rate (NMR). The latter is not the usual net migration to population ratio, but is defined in terms of the labor force in the previous time period, similar to the approach in some labor market studies. [21]

As in the previous section, the interest lies primarily in the illustration of spatial effects. The simple specification used here is not intended to make a contribution to the substantive understanding of the space—time dynamics of labor market adjustments. The model is estimated for observations on 25 South—Western Ohio counties in two time periods, 1983 and 1981. The spatial configuration of the counties is illustrated in Figure 2. The observations used in the empirical application are listed in Tables 12.16 and 12.17. These tables also include data on a dummy variable SMSA, which takes on the value of 1 for counties that are part of a metropolitan area (also illustrated in Figure 2). In addition, the time—lagged value for the dependent variable is listed, which is used as an instrument in the estimation in Section 12.3.3.

For the year 1983, the model is estimated for contemporaneous variables, i.e., WAG83 is regressed on UN83, NMR83, and SMSA. For 1981, this contemporaneous version achieved a very low explanatory power, and a lagged specification is used instead, with WAG80 and NMR80.

Spatial dependence is formalized in terms of first order contiguity, as given in Table 12.18. As before, all spatial weight matrices are used in row—standardized form.

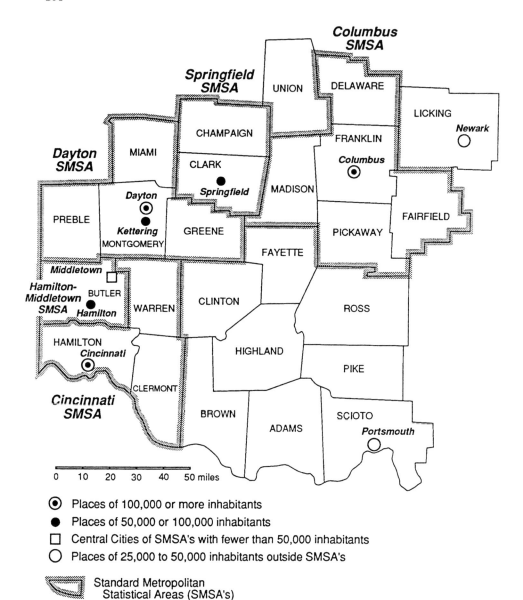

Figure 12.2.   Counties in South—Western Ohio.

Table 12.16. Observations for South–Western Ohio Counties–1983.

	COUNTY	WAGE83	UN83	NMR83	SMSA	WAGE82
1	UNION	1.003127	0.080500	−0.002217	0	1.108662
2	DELAWARE	1.039972	0.122174	0.018268	1	1.071271
3	LICKING	1.050196	0.095821	−0.013681	1	1.058375
4	MADISON	1.052210	0.102941	−0.000959	1	1.049791
5	FRANKLIN	1.055406	0.111965	−0.005160	1	1.075883
6	FAIRFIELD	1.048299	0.095192	−0.008934	1	1.056021
7	PICKAWAY	1.065595	0.100909	−0.017210	1	1.094584
8	CHAMPAIGN	1.035019	0.082500	−0.018383	1	1.099368
9	CLARK	1.059465	0.084819	−0.019710	1	1.044507
10	MIAMI	1.040293	0.086939	−0.023227	1	1.044117
11	PREBLE	1.025818	0.103500	0.001940	1	1.085981
12	MONTGOMERY	1.052522	0.095054	−0.023533	1	1.045559
13	GREENE	1.050858	0.109636	−0.030723	1	1.067446
14	BUTLER	1.018733	0.085530	−0.005622	1	1.032677
15	WARREN	1.057471	0.095000	−0.003058	1	1.061119
16	HAMILTON	1.052070	0.098056	−0.023413	1	1.070538
17	CLERMONT	1.027385	0.081975	0.000627	1	1.020961
18	CLINTON	1.034525	0.094500	−0.004150	0	1.052724
19	FAYETTE	1.022044	0.069444	−0.009074	0	1.065408
20	ROSS	1.023230	0.082571	0.018513	0	1.050953
21	HIGHLAND	1.057144	0.069545	0.003521	0	1.097586
22	BROWN	1.064552	0.061250	−0.004152	0	0.984695
23	ADAMS	0.932441	0.038182	0.023083	0	1.043908
24	PIKE	1.050463	0.060556	0.008225	0	1.088628
25	SCIOTO	1.032233	0.052500	−0.013592	0	1.063456

Table 12.17. Observations for South–Western Ohio Counties–1981.

	COUNTY	WAGE81	UN80	NMR80	SMSA	WAGE80
1	UNION	1.146178	0.130375	−0.010875	0	1.084886
2	DELAWARE	1.104241	0.189603	0.041886	1	1.110426
3	LICKING	1.094732	0.124125	−0.004158	1	1.069776
4	MADISON	1.065032	0.139394	−0.004271	1	1.067895
5	FRANKLIN	1.085075	0.183311	−0.015568	1	1.079768
6	FAIRFIELD	1.110068	0.155076	0.004514	1	1.099969
7	PICKAWAY	1.085111	0.138939	0.000000	1	1.067286
8	CHAMPAIGN	1.082043	0.133317	0.004439	1	1.114233
9	CLARK	1.046862	0.132806	−0.008948	1	1.070778
10	MIAMI	1.092917	0.129626	−0.021634	1	1.081339
11	PREBLE	1.066597	0.140589	0.019676	1	1.040860
12	MONTGOMERY	1.090118	0.129985	−0.004210	1	1.072204
13	GREENE	1.066719	0.155352	−0.001159	1	1.072529
14	BUTLER	1.099304	0.125243	−0.000649	1	1.090042
15	WARREN	1.083854	0.117648	0.002969	1	1.045758
16	HAMILTON	1.077636	0.149544	−0.028349	1	1.085611
17	CLERMONT	1.146295	0.114029	0.014909	1	1.141677
18	CLINTON	1.102698	0.122633	0.006075	0	1.079916
19	FAYETTE	1.063537	0.107812	0.023377	0	1.089024
20	ROSS	1.109413	0.118741	−0.015767	0	1.070656
21	HIGHLAND	1.062119	0.107926	−0.004996	0	1.063287
22	BROWN	1.060167	0.109170	−0.004790	0	0.936907
23	ADAMS	1.316997	0.065814	0.049736	0	1.059546
24	PIKE	1.143505	0.089019	0.052440	0	1.060121
25	SCIOTO	1.050139	0.073089	0.028002	0	1.048953

Table 12.18. First Order Contiguity for South—Western Ohio Counties

COUNTY   CONTIGUOUS TO:

County						
1:	2	4	5	8		
2:	1	3	5			
3:	2	5	6			
4:	1	5	7	8	9	13	19
5:	1	2	3	4	6	7
6:	3	5	7			
7:	4	5	6	19	20	
8:	1	4	9	10		
9:	4	8	10	12	13	
10:	8	9	12			
11:	12	14				
12:	9	10	11	13	14	15
13:	4	9	12	15	18	19
14:	11	12	15	16		
15:	12	13	14	16	17	18
16:	14	15	17			
17:	15	16	18	22		
18:	13	15	17	19	21	22
19:	4	7	13	18	20	21
20:	7	19	21	24		
21:	18	19	20	22	23	24
22:	17	18	21	23		
23:	21	22	24	25		
24:	20	21	23	25		
25:	23	24				

12.3.2. OLS Regression with Diagnostics for Spatial Effects

In order to set the stage for the later analyses, the Phillips−curve model is first estimated by OLS, and the presence of spatial effects is assessed by means of Lagrange Multiplier diagnostics. The estimation yields the following results, with the standard deviation and corresponding t−values listed in parentheses:

1983:

$$WAGE83 = \begin{array}{cccc} 0.967 & +0.960 \ UN83 & -1.027 \ NMR83 & -0.026 \ SMSA \\ (0.022) & (0.318) & (0.344) & (0.014) \\ (45.561) & (3.022) & (2.983) & (1.855) \end{array}$$

$$R^2 = 0.486 \qquad\qquad R_a^{\ 2} = 0.413$$

1981:

$$WAGE81 = \begin{array}{cccc} 1.163 & -0.674 \ UN80 & +1.059 \ NMR80 & +0.023 \ SMSA \\ (0.054) & (0.489) & (0.521) & (0.029) \\ (21.686) & (1.380) & (2.034) & (0.808) \end{array}$$

$$R^2 = 0.290 \qquad\qquad R_a^{\ 2} = 0.189$$

The results are clearly different for both years, in terms of parameter values, signs and significance, as well as in terms of overall fit. There is a strong indication of a Phillips−type effect in 1983, but for 1981 the coefficient of UN80 is not significant, and, moreover, it has the wrong sign. The effect of net migration is significant in both years, negative in 1983 and positive in 1981. The SMSA dummy is marginally significant and negative in 1983, but is not relevant in 1981.

The Lagrange Multiplier tests for spatial effects, listed in Table 12.19, provide a strong indication of misspecification, and slightly more so in 1981 than in 1983. The statistic shows a somewhat significant measure of spatial error autocorrelation in 1983, as well as a significant value for an omitted spatial lag. This is in contrast to the indication given by a Moran statistic for 1983 of $I=-0.263$, with an associated z−value of -1.563, which is not significant for p=0.10. A similar situation is present in 1981, where the Moran statistic of $I=-0.289$ and z−value of -1.868 are significant at p=0.06, whereas the LM statistic corresponds to a stricter critical level of 0.04. In both years there is also clear evidence of random coefficient variation. In order not to overburden the treatment in the remaining sections, the latter will be ignored, and the focus will be exclusively on various ways to incorporate spatial dependence.

12.3.3. IV Estimation of a Mixed Regressive Spatial Autoregressive Model

The indication of misspecification in the form of an omitted spatial lag motivates the inclusion of such a variable in the equation. In the context of a Phillips−curve, this gives a formal expression to a spatial spill−over effect in the wage ajustment, which is often cited in the literature as a consequence of

Table 12.19. Lagrange Multiplier Diagnostics for Spatial Effects

		1983		1981	
	q	χ^2	p	χ^2	p
Spatial Error Autocorrelation	1	3.574	0.06	4.295	0.04
Omitted Spatial Lag	1	4.718	0.03	5.429	0.02
Random Coefficient Variation	3	10.952	0.01	25.842	0.00

bargaining practices. The resulting mixed regressive spatial autoregressive model can be estimated by ML, as in the previous section, but also by instrumental variable techniques, as outlined in Section 7.1.2.

For comparison purposes, the ML results are listed in Table 12.20, which also includes the initial OLS estimate, the asymptotic standard deviation, and the associated t—value. The spatial autoregressive coefficient is strongly significant in both years, as indicated by a Wald test (asymptotic t test), with a probability of more than 0.99 in 1983, and 0.98 in 1981. A Likelihood Ratio test gives a similar indication: a value of 6.403 in 1983 (p=0.01) and 5.773 in 1981 (p=0.02). An LM statistic for spatial error autocorrelation is no longer significant. It yields a value of 1.630 in 1983, and 0.009 in 1981.

Instrumental variable estimation can be implemented in a number of different ways, depending on the choice of the instruments. To illustrate the sensitivity of the estimates to this choice, three cases are considered. One is the use of spatially lagged explanatory variables as instruments: W_UN83 and W_NMR83 for 1983, and W_UN80 and W_NMR80 for 1981. This is referred to as INST 1 in Table 12.21. below. The two other approaches, following a suggestion of Haining (1978a), consist of the use of a time—wise lagged dependent variable (INST 2), or its spatial lag (INST 3), i.e., WAGE82 or W_WAGE82 for the 1983 regression, and WAGE80 or W_WAGE80 for 1981.

A rough idea about the relative appropriateness of these instruments can be gained from their simple Pearson correlations with the spatially lagged dependent variable. In 1983, these are: 0.69 for W_UN83, −0.71 for W_NMR83, −0.08 for WAGE82, and 0.14 for W_WAGE82. In 1981, the correlations are: −0.59 for W_UN80, 0.80 for W_NMR80, −0.47 for WAGE80, and −0.05 for W_WAGE80. [22]

The resulting coefficient estimates are listed in Table 12.21, together with the asymptotic standard deviations. In all cases this standard deviation is larger than for the corresponding ML estimation, yielding a lower significance for the parameters. This also illustrates the extent of the inferior efficiency of the IV approach.

The IV—estimate for the spatial autoregressive coefficient varies greatly from one set of instruments to the other. It is marginally significant for INST 1 in 1983 (ρ=−0.547, p=0.14), and INST 3 in 1981 (ρ=−1.648, p=0.09), but for the latter the estimate is unstable (i.e., larger than one). This is also the case for INST 2 in 1981 ($\rho=-1.011$), although this estimate is not significant. For INST 2 in 1983, the resulting value for ρ even becomes positive (0.725), though not significant.

An improvement in efficiency may be obtained by taking a systems perspective, and allowing for inter−equation dependence via the error terms. Estimation is carried out by the Three Stage Least Squares approach (3SLS) outlined in Section 10.1.3. The resulting coefficient values are presented in Table 12.22, with all the spatially lagged explanatory variables (INST 1 for 1983 and 1981) used as instruments in both equations. [23] The 3SLS estimates have a smaller asymptotic variance, but the parameter significance for the spatial coefficient is not affected. It remains rather insignifiant: $\rho=-0.480$ with $p=0.14$ for 1983, and $\rho=-0.390$ with $p=0.23$ in 1981. On the other hand, both the unemployment (with $p<0.01$ in 1983, but $p=0.08$ in 1981) and the net migration variable (with $p<0.05$) are significant, similar to the result obtained in the ML estimation.

The systems approach allows for testing coefficient equality across equations. In the context of the space−time models, this is the same as testing for coefficient stability over time. For the spatial parameter, the corresponding Wald test yields a value of 0.055, for which the null hypothesis of equality cannot be rejected. However, it should be noted that the estimates for ρ were both not significantly different from zero in the first place.

Table 12.20. ML Estimation of the Spatial Phillips−Curve

1983: $L=75.392$

VARIABLE	OLS	ML	ST.DEV.	T
W_WAGE83	−0.848	−0.623	0.224	2.777
CONSTANT	1.834	1.604	0.230	6.965
UN83	1.035	1.015	0.246	4.126
NMR83	−0.834	−0.885	0.269	3.285
SMSA	−0.013	−0.017	0.011	1.444

1981: $L=54.166$

VARIABLE	OLS	ML	ST.DEV.	T
W_WAGE81	−0.801	−0.588	0.254	2.314
CONSTANT	2.069	1.827	0.292	6.250
UN80	−0.751	−0.731	0.387	1.891
NMR80	1.146	1.123	0.413	2.719
SMSA	0.0005	0.007	0.024	0.277

Table 12.21. IV Estimation of the Spatial Phillips—Curve

	INST 1		INST 2		INST 3	
1983						
VARIABLE	IV	ST.DEV.	IV	ST.DEV.	IV	ST.DEV.
W_WAGE83	−0.547	0.371	0.725	2.259	−0.120	3.007
CONSTANT	1.526	0.379	0.225	2.310	1.090	3.075
UN83	1.008	0.280	0.897	0.475	0.971	0.408
NMR83	−0.902	0.313	−1.192	0.695	−0.999	0.764
SMSA	−0.018	0.014	−0.037	0.039	−0.024	0.048
1981						
VARIABLE	IV	ST.DEV.	IV	ST.DEV.	IV	ST.DEV.
W_WAGE81	−0.290	0.387	−1.011	0.796	−1.648	0.978
CONSTANT	1.491	0.440	2.307	0.901	3.027	1.107
UN80	−0.702	0.458	−0.772	0.436	−0.833	0.517
NMR80	1.091	0.488	1.169	0.466	1.238	0.553
SMSA	0.015	0.029	−0.006	0.034	−0.024	0.041

Table 12.22. 3SLS Estimation of the Spatial Phillips—Curve

VARIABLE	3SLS	ST.DEV.	T
W_WAGE83	−0.480	0.326	1.474
CONSTANT83	1.464	0.334	4.385
UN83	0.902	0.246	3.666
NMR83	−0.699	0.269	2.602
SMSA83	−0.013	0.012	1.080
W_WAGE81	−0.390	0.323	1.207
CONSTANT81	1.606	0.367	4.371
UN80	−0.704	0.403	1.748
NMR80	0.901	0.420	2.144
SMSA81	0.008	0.026	0.327

$R^2 = 0.656$

12.3.4. SUR with Diagnostics for Spatial Effects

A systems perspective can also be taken towards estimation of the specification without the spatially lagged dependent variable. The allowance of inter—equation dependence via the error terms is incorporated in the familiar seemingly unrelated regression framework, discussed in Section 10.1.1.

The results for a SUR estimation of the Phillips—curve model in 1983 and 1981 are listed in Table 12.23. In addition to the parameter estimates, two measures of fit are presented. The first, R^2, is a pseudo R^2 in the form of the squared correlation between the observed and predicted values for the dependent variable. The second measure, R_B^2, is the R^2 adjusted for the form of the error variance, as suggested in Buse (1979), and is further discussed in Section 14.1.

Compared to the OLS values in Section 12.3.2, the SUR approach yields a smaller variance (as it should), but also leads to altered coefficient estimates. The combination of the two effects results in a lowered significance of the parameters for 1981. In particular, the coefficient for NMR81 is no longer significant at the 0.05 level (p=0.07). In 1983, both UN and NMR remain strongly significant.

Table 12.23. also lists the estimate for the inter—equation covariance matrix. A test on the significance of the implied temporal correlation can be based on the LM or LR principle, as outlined in Section 10.1.1. The corresponding statistics yield 5.658 for the LM test and 9.750 for the LR approach, both strongly significant (p=0.02 for LM, p<0.01 for LR). Consequently, there is a clear indication that the covariance matrix in question is not diagonal, and therefore that the SUR approach is relevant.

The Lagrange Multiplier test for the presence of spatial dependence in the error terms, which is derived in Section 10.1.4. yields a statistic of 3.779. With one degree of freedom, i.e., assuming equality of the spatial coefficient in both equations, this χ^2 variate is significant for p=0.05. However, this is no longer the case if two degrees of freedom are considered (i.e., a different coefficient in each equation).

12.3.5. Error Components with Diagnostics for Spatial Effects

An alternative perspective on spatial effects in space—time models is provided by the error components approach, outlined in Section 10.2.1. In this framework, spatial heterogeneity is incorporated in the error term, in the form of a space—specific variance $\sigma(\mu)$. In order to achieve this, the coefficients are typically assumed to be stable over time and across space. [24]

The results of a pooled OLS estimation of the model are presented in Table 12.24. Both UN and NMR turn out to be strongly significant (p<0.05), but SMSA is not. An indication of the need for the incorporation of error components is given by the Lagrange Multiplier test of Breusch and Pagan (1980). Its value of 9.433 is clearly significant, with p<0.01.

Maximum likelihood estimation of a 2ECM specification leads to the coefficients given in Table 12.25, with σ as the overall error variance, and $\omega=\sigma(\mu)/\sigma$ as the ratio of the space variance component to the overall variance, as in Section 10.2.3. As in the SUR model, the measures of fit are a pseudo R2 and the transformed R2 of Buse.

Table 12.23. SUR Estimation of the Phillips–Curve

VARIABLE	SUR	ST.DEV.	T
CONSTANT83	0.976	0.020	49.838
UN83	0.795	0.279	2.850
NMR83	−0.630	0.285	2.213
SMSA83	−0.016	0.012	1.283
CONSTANT81	1.163	0.047	24.677
UN80	−0.635	0.429	1.480
NMR80	0.793	0.431	1.840
SMSA81	0.017	0.026	0.661

$R^2 = 0.552$ $R_B^2 = 0.284$

Cross–Equation Error Covariance Matrix

$$\begin{matrix} 0.363\text{x}10-3 & -0.399\text{x}10-3 \\ -0.399\text{x}10-3 & 0.194\text{x}10-3 \end{matrix}$$

A test for residual spatial autocorrelation, based on the Lagrange Multiplier approach outlined in Section 10.2.3, yields a statistic of 0.916, which is clearly not significant (as χ^2 with one degree of freedom). In other words, after the introduction of spatial heterogeneity in the form of error components, there is no longer any evidence for spatial dependence in the error term. However, it should be noted that the 2ECM model achieves a fit which is considerably poorer than for the SUR approach, and therefore may not be the most relevant specification for the data set at hand.

Table 12.24. Pooled OLS Estimation of the Phillips–Curve

VARIABLE	OLS	ST.DEV.	T
CONSTANT	1.013	0.024	42.761
UN	0.584	0.253	2.307
NMR	0.954	0.403	2.370
SMSA	−0.010	0.018	0.519

$R^2 = 0.223$ $R_a^2 = 0.176$

Table 12.25. ML Estimation of the Two Error Components Model

VARIABLE	ML	ST.DEV.	T
CONSTANT	1.005	0.025	40.369
UN	0.676	0.260	2.604
NMR	0.983	0.433	2.272
SMSA	−0.012	0.021	0.604

Variance Components
$\sigma = 0.203 \times 10 - 2$
$\omega = 0.213$

$R^2 = 0.226$ $R_B^2 = 0.248$

12.3.6. Estimation of SUR with Spatial Dependence in the Error Terms

The final specification considered in this chapter consists of a SUR model with a different spatial dependence in the error term for each equation. The maximum likelihood approach to the estimation of this model was discussed in Section 10.1.2. It boils down to an iterated SUR on spatially transformed variables, in which the spatial autoregressive coefficients are obtained from a maximization of a concentrated likelihood.

The starting values for the iterative procedure are obtained from a ML estimation for each equation separately, which yields $\lambda_1 = -0.950$ and $\lambda_2 = -0.854$. A complete list of the values for the λ at each iteration is presented in Table 12.26.[26]

The ML estimates are listed in Table 12.27. Compared to the standard SUR in Table 12.23, they achieve a substantial reduction in the parameter variance, which results in higher (asymptotic) significance. In fact, NMR is now strongly significant in both years, and SMSA becomes significant for 1983. As before, UN is significant only in 1983, but at a much higher level.

The spatial coefficients are strongly significant as well, as indicated by the asymptotic t−values listed in the table (p<0.01). A Wald test on the joint significance of λ_1 and λ_2 yields a value of 32.776, significant for a p<0.01. Similarly, a Likelihood Ratio statistic gives a result of 5.995, which is significant at p=0.05. Since the LM test from Section 12.3.4. led to a statistic of 3.779, this is an instance where the small sample inequality between the asymptotic tests (LM<LR<W) would provide conflicting indications. As pointed out earlier, the LM statistic for a null hypothesis of a different spatial dependence in each equation is not significant at p=0.05, whereas the LR and W reported here clearly are.

A more explicit Wald test on the null hypothesis that $\lambda_1 = \lambda_2$ can be derived from the variance and covariance between these estimates, as outlined in Section 6.3.1. The relevant variances are 0.0262 for λ_1, 0.0403 for λ_2, and the covariance is 0.0325. The difference $\lambda_1 - \lambda_2 = -0.142$. Consequently, a Wald statistic for this hypothesis is found as $(-0.142)^2/0.0015 = 13.236$ (rounded from a more precise calculation), where the denominator is obtained from the quadratic form in the variance matrix, and in a vector which corresponds to the coefficient constraint $[1 \ -1]$. This value is clearly significant at $p < 0.01$, which turns out to provide an a posteriori justification for the choice of this model specification.

Table 12.26. Iterations for the Spatial Coefficients in the ML
Estimation of the Spatial SUR Model

ITERATION	λ_1	λ_2
1	−0.950	−0.854
2	−0.932	−0.836
3	−0.939	−0.821
4	−0.941	−0.813
5	−0.943	−0.808
6	−0.943	−0.806
7	−0.944	−0.804
8	−0.944	−0.803
9	−0.944	−0.803
10	−0.944	−0.802
11	−0.944	−0.802

Table 12.27. ML Estimates for the SUR Model with Spatial Error
Autocorrelation

VARIABLE	ML	ST.DEV.	T
CONSTANT83	0.974	0.0002	77.388
UN83	0.853	0.036	4.799
NMR83	−0.689	0.053	3.549
SMSA83	−0.020	0.0001	2.857
CONSTANT81	1.136	0.001	32.692
UN80	−0.374	0.104	1.094
NMR80	1.179	0.161	3.372
SMSA81	0.007	0.0003	0.547
λ_1	−0.944	0.162	5.838
λ_2	−0.802	0.201	3.995

$L = 159.434$ $\qquad$ $R^2 = 0.547$

Appendix 12.A.   **A Bisection Search for the Coefficient in a Pure Spatial Autoregressive Model** [27]

A bisection search procedure includes the optimal value for the parameter to the desired degree of accuracy between a lower (L) and upper bound (U), for which:

$$\partial L_C / \partial \rho (\rho_L) < 0$$

$$\partial L_C / \partial \rho (\rho_U) > 0$$

where L_C is the concentrated likelihood, and ρ is the spatial parameter. This optimal value is obtained by successively taking the midpoint between an upper and lower bound. This midpoint in turn becomes an upper or lower bound, depending on the value for the partial derivative.

In a pure first order spatial autoregressive model, the OLS estimate for the spatial parameter and the Moran coefficient provide natural starting values. The latter corresponds to a regression of a spatial lag on the original variable and is always smaller than the OLS estimate in absolute value. Moreover, both OLS and Moran estimates always have the same sign. [28]

The designation of the starting lower and upper bounds proceeds as follows.

[1]   check the value for the OLS estimate: if larger than one in absolute value, set ρ_1 equal to $e.|1-\delta|$, where e is $+1$ or -1 such that ρ_1 has the same sign as the OLS estimate, and where δ is a small value such that the Jacobian evaluated at $|1-\delta|$ is nonsingular; set ρ_1 equal to the OLS estimate if it is less than one in absolute value

[2]   set ρ_2 equal to the Moran coefficient

[3]   compute $\partial L_C / \partial \rho$ for ρ_1 and ρ_2

[4]   if $\partial L_C / \partial \rho(\rho_1).\partial L_C / \partial \rho(\rho_2) < 0$ and $\partial L_C / \partial \rho(\rho_1) > 0$, then $\rho_L = \rho_2$, and $\rho_U = \rho_1$; else $\rho_L = \rho_1$, and $\rho_U = \rho_2$; proceed to the bisection routine

[5]   if $\partial L_C / \partial \rho(\rho_1) > 0$ and $\partial L_C / \partial \rho(\rho_1) > \partial L_C / \partial \rho(\rho_2)$, then $\rho_L = -(1-\delta)$ and $\rho_U = \rho_2$; proceed to the bisection routine

[6]   if $\partial L_C / \partial \rho(\rho_1) > 0$ and $\partial L_C / \partial \rho(\rho_1) < \partial L_C / \partial \rho(\rho_2)$, then $\rho_L = \rho_1$ and $\rho_U = 1-\delta$; proceed to the bisection routine

[7]   if $\partial L_C / \partial \rho(\rho_1) < 0$ and $\partial L_C / \partial \rho(\rho_1) > \partial L_C / \partial \rho(\rho_2)$, then $\rho_L = \rho_1$ and $\rho_U = 1-\delta$; proceed to the bisection routine

[8]   if $\partial L_C / \partial \rho(\rho_1) < 0$ and $\partial L_C / \partial \rho(\rho_1) < \partial L_C / \partial \rho(\rho_2)$, then $\rho_L = -(1-\delta)$ and $\rho_U = \rho_2$; proceed to the bisection routine.

The bisection routine takes the midpoint as $\rho_M = (\rho_L + \rho_U)/2$. This midpoint becomes the new lower bound if the partial derivative of the concentrated likelihood is negative, and the new upper bound otherwise. This process continues until the desired precision is obtained. The latter is equal to the difference between the upper and lower bound.

NOTES ON CHAPTER 12

[1]   Although OLS on the transformed variable is easily carried out, the iteration back and forth between the estimation of β and the optimization for λ may be tedious in an econometric package without macro facilities.

[2]   Typically, a simultaneous optimization of the likelihood for all parameters will be numerically more complex. Moreover, not all standard optimization routines will converge to coefficient values within an acceptable range, i.e., spatial parameters less than one in absolute value and positive estimates for variance terms. Also, a number of approaches are very sensitive to the proper choice of a starting value for the parameters, and may not converge at all if this is not achieved. A further discussion of these issues is given in Section 12.1.4.

[3]   It is important to note that a standardized weight matrix will not be symmetric, so that a more complex numerical procedure is needed to compute the eigenvalues. The most common routines are based on the so-called QR algorithm. For details, see, e.g., the algorithms discussed in Wilkinson and Reinsch (1971). These routines are present in mathematical packages such as IMSL and LINPACK, but are not always included in standard statistical packages, where often only the symmetric case is considered. Ord (1975) suggests a transformation of the nonsymmetric matrix to a symmetric one which has the same eigenvalues (but not the same eigenvectors).

[4]   As shown in Chapter 6, the partial derivative of $\ln|A|$ with respect to ρ is $-\mathrm{tr}\,A^{-1}W$.

[5]   For example, all illustrations in the following sections are computed by means of routines written in the GAUSS matrix language, which is optimized for precision and speed for matrices smaller than 90 by 90. See Edlefsen and Jones (1986) for a description of the language.

[6]   It should be noted that a smaller scale factor will slow down the speed of convergence, since the parameter moves towards the local optimum by smaller increments.

[7]   See also Anselin and Griffith (1988) for an elaboration of this point.

[8]   Examples of the integration of spatial routines into SAS and MINITAB are presented in Griffith (1988b). Instrumental variables estimators for spatial process models are incorporated into SAS in the routines reported in Anselin (1985).

[9]   Some examples are the FORTRAN routines for measures of spatial association based on the Hubert—Golledge quadratic assignment approach, in Costanzo (1982) and Anselin (1986c), and a set of FORTRAN programs for the estimation of spatial process models listed in Anselin (1985). Other programming efforts are sometimes referred to in the footnotes of journal articles.

[10]   The traces needed in the computation of the variance matrix are: $\mathrm{tr}(WA^{-1})^2=19.440$ and $\mathrm{tr}(WA^{-1})'(WA^{-1})=22.338$.

[11]   The values listed in Table 12.5. are rounded. The actual estimation was carried out with a convergence criterion associated with a precision of better than six decimals. The

corresponding estimate is 0.431023, with a derivative of the concentrated likelihood of 0.0000016, obtained after 23 iterations.

[12] See, e.g., the discussion in Anselin (1980).

[13] As before, the values reported here are rounded. The final value is rounded from a more precise calculation, and is not the result of a manipulation of the rounded values.

[14] The traces necessary in the computation of the variance matrix are: $tr(WB^{-1})^2=27.551$, and $tr(WB^{-1})'(WB^{-1})=31.562$.

[15] The values in the table are rounded. With a more rigorous precision criterion of six decimals, the estimation takes 11 iterations. The final estimate is $\lambda=0.561790$, with an associated σ^2 of 95.5745.

[16] A further analysis of the common factor hypothesis is given in Section 13.3.1.

[17] Estimation was carried out by means of and iterated EGLS, with an adjusted Gauss–Newton procedure to obtain estimates for the random components. Starting values were generated from a Godfeld–Quandt approach. The latter consists of regressing the squared residuals on the squared explanatory variables. Convergence, with a precision criterion of six decimals, was achieved in 25 rounds for the model with INC and HOUSE, and in 9 rounds for the model with HOUSE only.

[18] The corresponding Moran statistics for the residuals are $I=0.256$ and $I=0.208$, with associated z–values of 2.995 and 2.472 (under the assumption of randomization).

[19] Critical levels for an $F(3,43)$ variate are 2.83 for p=0.05 and 4.28 for p=0.01.

[20] As before, the results in Table 12.14. are rounded. The actual estimation was carried out with a precision criterion of 6 decimals and converged after 15 iterations. The final values of the parameters were: $\lambda=0.64803$, and for the random components: CONSTANT=67.27923, HOUSE=0.00991.

[21] For a recent example, see, e.g., Greenwood, Hunt and McDowell (1986).

[22] A more rigorous approach could be based on canonical correlations, as in Bowden and Turkington (1984). See also the remarks in Section 7.1.1.

[23] The measure of fit in the 3SLS results is a pseudo R^2, in the form of the squared correlation between observed and predicted dependent variables.

[24] Strictly speaking, for the example used here this approach is not appropriate, since it ignores the different time lags involved in the explanatory variables. Aside from this conceptual problem, which is less important in this purely numerical illustration, there is statistical evidence that the coefficients are not stable over time. For example, for the SUR estimates in Table 12.23, a Wald test on coefficient equality rejects the null hypothesis for both UN (W=5.770, p=0.02) and NMR (W=6.289, p=0.01) individually, as well as for all parameters jointly (W=16.460, p<0.01).

[25] Estimation is carried out by iterated EGLS, where the estimates for σ and ω at each iteration are based on transformed residuals, in the standard way. Convergence, with a precision of more than six decimals, is reached after 7 iterations.

[26] As before, the values in Table 12.26. are rounded, and the actual estimation was carried out with a precision of higher than 6 decimals. This converged after 22 iterations and yielded the estimates $\lambda_1 = -0.944042$ and $\lambda_2 = -0.801991$.

[27] An earlier version of this algorithm is presented in Anselin (1980, pp. 50-53).

[28] In a mixed regressive spatial autoregressive model there is no equivalent to a Moran coefficient. A suitable alternative may be to take zero, or a fraction of the OLS estimate. In the illustrations presented here, a fraction of 0.2 is used. Even though it is chosen rather arbitrarily, this value has provided a useful starting point in many other empirical implementations as well.

PART III

MODEL VALIDATION

CHAPTER 13

MODEL VALIDATION AND SPECIFICATION TESTS

IN SPATIAL ECONOMETRIC MODELS

In philosophical discussions about the nature of progress in science it is often maintained that the development of theory cannot proceed without some form of reality—based validation. This leads to a continual process of interaction between theory formation and empirical assessment, motivated in part by the observation of phenomena which are not explained by existing theories, and by the failure of theoretical constructs to be reflected in experimental situations. Essential to this process is a set of clear standards or criteria of validity, and a uniform methodology to apply these to the theoretical constructs under scrutiny. Typically, this is based on a formal probabilistic framework.

In this and the next chapter, I focus on econometric techniques to assess the validity of spatial models in regional science. [1] This issue is increasingly relevant in applied modeling, as competing theories proliferate and policy decisions become more and more based on quantitative projections and impact assessments. From a methodological viewpoint as well, there is a growing awareness in spatial econometrics of the need to more effectively deal with the empirical validation of spatial theoretical constructs, similar to the expression of general concerns about the relevance of statistical modeling in the other social sciences. [2]

In the literature of regional science, economic geography and planning theory, the appropriateness of model building and statistical analysis itself is sometimes questioned. For example, in Gould (1970, 1981) and Fischer (1984), this is based on philosophical considerations, while in Alonso (1968) and Lee (1973) more operational concerns are voiced. More recently, such concerns have resulted in considerable attention to the formal evaluation of the performance of regional and multiregional econometric models, as in Taylor (1982a, 1982b), Charney and Taylor (1983, 1984), and in the development and application to spatial analysis of a variety of econometric model validation techniques, e.g., in Buck and Hakim (1980, 1981), Burridge (1981), Horowitz (1982, 1983, 1985, 1987), Blommestein (1983), Bivand (1984), Anselin (1984a, 1984b, 1984c, 1986a, 1987a), Anselin and Can (1986), and Blommestein and Nijkamp (1986).

The general problem of model validation, as it applies to spatial aspects of models in regional science is considered in more detail in this chapter. First, I will discuss in broad terms the range of issues relevant to model validation in spatial econometrics. This is followed by a more extensive treatment of model specification tests, first in general and then more in detail, for tests on common factors and tests on non—nested hypotheses. Model selection and model discrimination are considered in Chapter 14.

As in previous chapters, the emphasis is on techniques that address problems related to the spatial organization of data, or the spatial expression of a theory. Some general aspects of model validation, such as tests for functional misspecification, the selection of regressors, and the evaluation of predictive performance are typically not affected by the spatial nature of the analysis. Since these techniques can therefore largely be applied in the usual fashion, they will not be considered here.

13.1. General Issues of Model Validation in Spatial Analysis

It is generally accepted that the state of the art in spatial theory is still far from adequate to deal with the full scope of real problems faced by cities and regions. In this respect, larger and more complex models and theories have not always contributed to our knowledge base in proportion to their size. Indeed, general principles of scientific research, such as the search for parsimony, Occam's razor and other simplicity postulates are often ignored. For example, as argued in Jeffreys (1967), Gould (1981), and Zellner (1984), these principles should provide important guidance in the quest for superior formulations. In regional science and geography, it is often pointed out that a greater emphasis on precise axiomatic constructs, the use of more direct behaviorally inspired relations, and an attention to empirical regularities could lead to major improvements in this regard. [3]

Consequently, the spatial theories to which the statistical model validation pertains are often rather simple statements. Therefore, as a general caveat for the discussion which follows in this and the next chapter, it should be kept in mind that carrying out a rigorous analysis on this weak basis may not always be a very meaningful exercise.

In addition, from an empirical standpoint, the lack of controlled experimentation in the analysis of human spatial behavior limits the extent to which observed relationships can be attributed to general patterns and structures. As is well known, this is accompanied by a host of measurement problems. In spatial analysis, the poor quality of data collected for aggregate spatial units of observation is notorious. The resulting problems associated with spatial dependence and spatial heterogeneity were discussed at length in earlier chapters, particularly in Part II.

In the context of model validation, a general problem is the existence of potentially conflicting indications obtained from different criteria. In this respect, it is important to make the distinction between two different categories of approaches: specification tests and model selection techniques. While the former are embedded in a hypothesis testing framework, and thus are part of standard statistical inference, the latter are not formulated as hypotheses. They are often based on measures of fit and more ad hoc decision rules.

In general terms, model specification tests (or misspecification tests) consist of procedures to assess whether the deterministic or stochastic specifications of a model are borne out by the data. These tests include diagnostic checks on the presence of ways in which a particular formulation may be defective, such as tests for nonlinearities, or tests for nonspherical error terms, as well as tests on hypotheses about the validity of alternative specifications, such as in tests on non-nested hypotheses. In contrast, model selection and model discrimination techniques are decision rules which allow a researcher to select one from among a number of competing formulations, based on specific indicators. [4]

Since the distinction between specification tests and model selection is generally accepted in the econometric literature, it will also form the basis for the structure of this and the next chapter. However, it should be noted that this distinction is not always clearcut and without ambiguities. For example, to carry out a hypothesis test in a classical (non−Bayesian) approach necessitates the assumption of a true null hypothesis, and therefore logically precludes the consideration of other models as potentially correct. However, in practice this requirement is often ignored, as pointed out in various discussions of data mining and other specification searches, e.g., in Leamer (1978) and Lovell (1983). Another potential source of confusion about the two approaches lies in the property that some indicators of fit used in model selection can be related to hypothesis tests. For example, this is the case for the well−known R^2 measure of goodness of fit and the F test on the significance of a vector of regression coefficients.

From an operational viewpoint, the results of specification tests and of model selection procedures need to be incorporated in the decision process of the modeler. Therefore, conceptually, both approaches should be taken to form part of an overall and potentially qualitative model validation framework. Indeed, in applied work, given the variety and lack of precision of regional science theories, quite often several competing models can be estimated on the same data set (e.g., different spatial interaction models, urban density functions, economic base specifications).

The indicators provided by a battery of specification tests and measures of fit that are routinely generated by most econometric packages have to be consolidated in a consistent decision framework. This framework should formally take into account the multiplicity of criteria and the potentially subjective nature of the decision process of the individual analyst. This issue is largely ignored in the spatial (and, to a lesser extent, the non−spatial) econometric literature, but needs to be addressed in order to maintain the credibility and internal consistency of the methodological apparatus.

13.2. Specification Testing in Spatial Econometrics

In spatial analysis, three general types of specification tests are particularly relevant. The first concerns tests for the presence of spatial effects, i.e., tests for spatial dependence and spatial heterogeneity. These have been discussed at length in previous chapters (Chapters 6, 8 and 9) and will not be further considered here.

The second class of tests pertains to the extent of spatial effects, in particular, the extent of spatial dependence as reflected in the lag length included in spatial process models. This leads to the spatial common factor approach, discussed in Section 13.3.

The third class of tests is relevant in determining the structure of spatial dependence, as reflected in the choice of a spatial weight matrix. Tests on non−nested hypotheses are particularly appropriate in this context. They are considered in more detail in Section 13.4.

In more formal terms, the three types of tests can be illustrated by means of the following general specification:

$$y = g(y,\rho) + X\beta + \epsilon$$
$$\epsilon = h(\epsilon,\lambda) + \mu$$

where, as before, y is a N by 1 vector of values for a dependent variable, observed across spatial units, $g(y,\rho)$ is a function which expresses the spatial dependence among the y, with parameters ρ, X is a set of exogenous explanatory variables with associated parameters β, and ϵ is a disturbance term. The disturbance term itself is potentially spatially structured as well, formalized in the function $h(\epsilon,\lambda)$, with parameters λ.

Typically, both functions g and h are linear weighted sums, obtained from the multiplication of a spatial weight matrix W with the vector of observations on y (or ϵ).

The first type of specification test, on the existence of spatial effects, consists of determining whether the $g(y,\rho)$ or $h(\epsilon,\lambda)$ are relevant. The second type of test, on common factors, addresses the length of the spatial lag, i.e., the spatial extent of the dependencies encompassed in $g(y,\rho)$ or $h(\epsilon,\lambda)$. In other words, this test attempts to find how many powers of W should be included in the functional specification. The third type of test, on non−nested hypotheses, investigates the structure of spatial dependence incorporated in $g(y,\rho)$ or $h(\epsilon,\lambda)$, i.e., the structure of the W itself.

13.3. Determination of Spatial Lag Length: Tests on Common Factors

Following the work on dynamic specification analysis of econometric models in a time series context, e.g., by Zellner and Palm (1974), Hendry and Mizon (1978), Mizon and Hendry (1980), Sargan (1980), and Kiviet (1985, 1986), tests on common factors have received recent attention in spatial econometrics, e.g., in Burridge (1981), Blommestein (1983), Bivand (1984), and Blommestein and Nijkamp (1986).

In this section, I will briefly outline the general principle behind the spatial common factor approach, and discuss its relevance for specification analysis in applied spatial modeling.

13.3.1. General Principle of the Spatial Common Factor Approach

In time series analysis, the common factor approach is based on the equivalence of two model specifications, one expressed in terms of the errors, the other in terms of lagged dependent variables. In essence, if one or more common coefficients are present for different lags of the dependent variables, the specification reduces to a simpler form with lagged errors.

In spatial analysis, this situation is similar to the equivalence between a model with a spatially autoregressive error term and the spatial Durbin approach, as illustrated in Section 8.2.2.

The least complex situation is the one considered by Burridge (1981) and Bivand (1984), where only one spatial lag is taken into account. The starting point is a the linear model with a spatially autoregressive disturbance, in the usual notation, expressed as:

$$y = X\beta + \epsilon$$
$$\epsilon = \lambda.W.\epsilon + \mu$$

or, as before:

$$y = X\beta + (I - \lambda.W)^{-1}\mu.$$

After pre−multiplying both sides of this equation by $(I-\lambda W)$, a mixed spatially regressive/autoregressive model results, as in the spatial Durbin approach:

$$y = \lambda Wy + X\beta - \lambda WX\beta + \mu,$$

or

$$y = \lambda Wy + X\beta + WX\gamma + \mu.$$

Reversing the flow of reasoning, the more complex autoregressive model can be reduced to a simple regression of y on X with a spatially autoregressive disturbance, if certain constraints are satisfied on the coefficients of the more complex model. More precisely, in order for the two specifications to be equivalent, the product of the coefficients of Wy and X (λ times β) should equal the negative of the coefficient of WX ($\lambda.\beta$, or $-\gamma$). In other words, a total of K nonlinear constraints need to be satisfied, one for each element of the vector β. However, when a row−standardized weight matrix is used, the two constant terms cannot be separately identified, and thus only K−1 constraints can be verified.

In contrast to the situation in time series analysis, the estimation of the simple regression model with spatially dependent errors is not easier than for the model with spatially lagged dependent variables. In both instances, maximum likelihood is the preferred approach, as indicated in Chapter 6. However, as pointed out in Section 12.1, the estimation of the model with spatially dependent errors is numerically more complex. The only potential advantage of this specification lies in its smaller number of parameters, i.e., K+1 instead of 2K + 1, which should result in improved efficiency (less multicolinearity).

Based on ML estimation, the test on the coefficient constraints is carried out by means of an asymptotic Likelihood Ratio, Wald or Lagrange Multiplier statistic, distributed as χ^2 with K−1 degrees of freedom (ignoring the constant terms). Formally, in the same notation as above, the null hypothesis is:

$$H_0 : \lambda.\beta + \gamma = 0$$

which, ignoring the constant term, consists of $K-1$ constraints on the $2K + 1$ parameters in the model (not including σ^2). The various tests can be carried out as straightforward applications of the general testing approach outlined in Section 6.3.1. For example, for a Wald test, an auxiliary matrix is needed of the partial derivatives of the constraints with respect to the model parameters:

$$G = \partial g'/\partial \theta$$

where

$$
\begin{aligned}
g &= \lambda\beta + \gamma \\
\theta' &= [\lambda \ \beta' \ \gamma']
\end{aligned}
$$

and β does not include the constant term.

From a simple application of matrix derivatives, the relevant elements of G follow as:

$$
\begin{aligned}
\partial g/\partial \lambda &= \beta \\
\partial g/\partial \beta &= \lambda I \\
\partial g/\partial \gamma &= I
\end{aligned}
$$

Consequently, as in Chapter 6, the corresponding Wald test is of the form:

$$W = g'[G'VG]^{-1}g \sim \chi^2(K-1)$$

with V as the estimated coefficient covariance matrix. The LR and LM tests can be developed in similar ways.

As an illustration, consider the spatial Durbin model estimated in Section 12.2.5. From the second column in Table 12.8, the ML estimates for the coefficients are:

$$\lambda = 0.426$$

$$\beta' = [-0.914 \quad -0.294]$$

$$\gamma' = [-0.520 \quad 0.246]$$

The corresponding vector g is:

$$g' = [-0.910 \quad 0.120]$$

This leads to a value of the Wald statistic of 4.228, which is not significant at $p=0.10$ for a χ^2 variate with 2 degrees of freedom. In other words, the common factor hypothesis cannot be rejected, which indicates the validity of a specification with a spatially autoregressive error term.

The common factor approach can be extended to several degrees of spatial lags, i.e., to models with higher powers of the spatial weight matrix W, as well as to various time−space lags, as shown by Blommestein (1983) and Blommestein and Nijkamp (1986). They refer to this method by the acronym COMFAC.

A complication arises in instances where the complexity of the different lag structures leads to restricted models that are difficult to interpret. In order to structure the ensuing specification search in accordance with a priori insights from economic or spatial theory, various additional constraints may be imposed. This variant of the COMFAC approach is referred to as ECONFAC. As Blommestein (1983) points out, the a priori constraints should be based on theory. For example, this could consist of coefficient restrictions derived from a rational expectations model, or constraints related to spatial structure.

A practical implementation of the common factor approach consists of a sequential procedure in which a general, overparameterized model with a large lag length is subjected to a series of tests on coefficient constraints for subsets of the parameters. In other words, rather than starting from a simple specification and proceeding towards more complex models, the common factor approach takes the other direction. In this way, misspecification from ignoring potentially relevant larger lags can be avoided. [5]

Interestingly, Burridge, Blommestein and Bivand all apply the common factor idea to the same data set, but obtain slightly different results. [6] The differences seem to be attributable not only to the relevance of different lag structures, as the common factors approach is intended to assess, but also to the specification of the spatial weight matix W. For example, by using a standardized binary contiguity matrix, Bivand eliminates residual spatial autocorrelation which was reported by the other authors. [7] It should be noted that in the common factor approach the spatial weight matrix is taken as given, and that its appropriateness is not subject to scrutiny.

13.3.2. Evaluation and Practical Implications

Although the common factor approach provides a formal basis for a structured specification analysis of spatial process models, a number of concerns may limit its usefulness in an applied context. Since the primary focus is on the spatial lag length, the structure of the spatial weight matrix itself is considered to be known. Consequently, an important aspect of the specification analysis may be ignored. Moreover, the choice of a different (*wrong*) W matrix can lead to a misleading interpretation of the significance of the lag length.

In addition, the issue of the proper spatial lag length may not be very practical in applied work. Since most spatial data situations are characterized by a small number of observations, any realistically sized analysis which focuses on several spatial lags will tend to quickly run out of degrees of freedom. This will also negatively affect the efficiency of the estimates on which the determination of the common factor constraints is based.

From a more formal perspective, the sequential nature of the tests on coefficient constraints may lead to problems of interpretation in a classical probabilistic framework. To assess the proper marginal significance levels associated with these multiple comparisons, so−called Bonferroni bounds could be applied (Savin 1980), which roughly consist of dividing the overall desired significance level by the number of comparisons carried out. This implies that, in order to remain consistent within the overall framework, in each of the marginal tests a much lower significance level than the traditional 0.05 or 0.10 should be applied. As a result, stricter standards will be used for the significance of an individual parameter, and the null hypothesis will tend to be rejected in fewer instances. It is therefore important not to carry out a mechanical interpretation of significance levels in the usual fashion, since this does not retain its internal validity in the case of multiple comparisons. [8]

The distributional characteristics of estimators which are found after a sequence of specification tests do not correspond to the usual asymptotic results. In essence, the coefficients are obtained as pre–test estimators, for which the distributions are typically highly complex, and not easily expressed in analytic terms, as pointed out in Section 11.1.

Finally, as noted before, the asymptotic significance levels for the LR, W and LM test statistics used in the common factor approach may be misleading in situations with few observations. As much as possible, finite sample corrections should be implemented, although this will considerably increase the numerical complexity of the various test. Indeed, the multilateral pattern of dependence associated with spatial autoregressive formulations in higher lags precludes the direct application of standard approximations, such as the procedures outlined in Rothenberg (1984b). Consequently, the implementation of a common factor approach in realistic modeling situations may not always be very practical.

13.4. Determination of Spatial Structure: Non–Nested Tests

In many situations in applied regression analysis, competing models can be formulated as alternative hypotheses. When one model or hypothesis cannot be expressed as a special case of the other, the hypotheses are considered to be non–nested, and special test procedures are needed.

The application of tests on non–nested hypotheses in spatial analysis is rather recent, although it was suggested as a potentially fruitful approach in the *Spatial Econometrics* book by Paelinck and Klaassen (1979). More recent overviews of the characteristics of this type of tests and its usefulness in regional and urban analysis are presented in Horowitz (1982, 1983, 1987), and Anselin (1984a, 1984c, 1987a). Increasingly, non–nested tests have also been applied in empirical studies in regional science, for example, in the specification of alternative discrete choice models in transportation analysis by Horowitz (1983, 1987), in the evaluation of aggregate spatial interaction functions by Anselin (1984c), and in the comparison of urban density models by Anselin and Can (1986).

In this section, I will consider the way in which the non–nested approach is applied in specification tests for the structure of spatial dependence in the weight matrix of spatial process models. Clearly, the lack of an overall encompassing framework for the weights that are incorporated in spatially lagged variables necessitates the use of a non–nested approach when comparing competing specifications.

After a brief formal introduction and review of various approaches that have been suggested in the literature, I focus more specifically on tests based on ML estimation, and tests based on IV estimation. The section is concluded with an evaluation of the practical relevance of these techniques in empirical work in regional science. [9]

13.4.1. General Principle of Tests on Non-Nested Hypotheses

Tests on non-nested hypotheses are designed to deal with the situation where competing model formulations cannot be considered as limiting forms of a more general expression. In other words, hypotheses are considered as non-nested when there is no natural encompassing framework, in which each can be derived as a special case. This is commonly encountered in applied work, for example, when both linear or multiplicative functional forms could be considered. This may also occur when conflicting conceptual frameworks preclude the joint inclusion of explanatory variables, such as demand-oriented or supply-side theories. Similarly, when several distributional assumptions are equally valid, such as in the use of logit or probit in discrete choice modeling, the specification tests are of a non-nested nature. In spatial analysis, this type of conflict typically surfaces in the consideration of different formulations of spatial friction, spatial dependence, or spatial structure.

In general, non-nested tests can be considered as part of a unified framework to carry out the formal comparison of explanatory variables, functional forms, and combinations of the two. The tests have become fairly well-known in econometric work, as illustrated in the comprehensive reviews of MacKinnon (1983) and McAleer (1987). Early approaches were based on the statistical work of Cox (1961, 1962) and Atkinson (1970). More recently, several extensions of these ideas have been developed for many situations encountered in regression analysis, among others, by Pesaran (1974, 1981, 1982a), Quandt (1974), Pesaran and Deaton (1978), Fisher and McAleer (1981), Davidson and MacKinnon (1981, 1982, 1984), White (1982b), Aguirre-Torres and Gallant (1983), MacKinnon, White, and Davidson (1983), Ericsson (1983), and Godfrey (1983, 1984). [10]

In formal terms, the alternative specifications that make up the non-nested hypotheses are considered in pairwise fashion. For example, for the case of two linear spatial process models, this can be expressed as:

$$H_i: y = \rho_i W_i y + X_i \beta_i + \epsilon_i$$

$$H_j: y = \rho_j W_j y + X_j \beta_j + \epsilon_j$$

In these expressions, y is the dependent variable of interest (or a transformation thereof), the $W_{i(j)}$ are different a priori structures for spatial dependence in the weight matrix, the $X_{i(j)}$ are sets of explanatory variables, some of which may be common to both expressions, the $\rho_{i(j)}$ and $\beta_{i(j)}$ are model coefficients, and the $\epsilon_{i(j)}$ are disturbance terms that do not necessarily have the same distribution.

Non-nested tests are carried out by considering each specification in turn as an alternative hypothesis. For example, the model represented by H_j would be considered in an attempt to falsify the maintained formulation H_i. The tests are non-symmetric in the sense that the falsification of H_i by H_j does not preclude H_j from being falsified by H_i.

Several different test procedures have been suggested, most of which boil down to a form of modified Likelihood Ratio test, or to a test on the significance of the nesting parameter in an artificial encompassing model.

The first type of approach is reflected in the well–known Cox–test and its extensions, such as the CPD test of Pesaran and Deaton (1978). The test consists of the comparison of two likelihood ratios. The first is constructed under the maintained hypothesis for each model, i.e., assuming each model is correct. The second ratio is based on the assumption that only one of the specifications is valid, namely the one associated with the null hypothesis. Consequently, it involves the estimation of the second model when the first is assumed correct. This is achieved by means of a pseudo maximum likelihood approach. Similar tests, based on instrumental variables estimation, such as the T and G tests of Godfrey (1983) and Ericsson (1983), are constructed from quadratic forms in the residuals.

In the artificial nesting approach, a manipulation of predicted values of the second model is included as an additional explanatory variable in an augmented first model. Lack of significance of the nesting parameter points to the strength of the first model in explaining the data. In other words, if no additional explanatory power is provided by including a manipulation of the second specification into the first, this is considered as evidence for the appropriateness of the maintained hypothesis. This idea is used in the J, P, and P_E tests of Davidson and MacKinnon (1981), and MacKinnon, White and Davidson (1983). The main differences between the various tests arise from the use of distinct asymptotic approximations to obtain consistent estimates.

Next, I outline in some more detail how these types of tests can be applied to specification testing in spatial process models.

13.4.2. Non–Nested Tests on the Spatial Weight Matrix Based on Maximum Likelihood Estimation

Maximum likelihood estimation forms a natural point of departure for many tests on non–nested hypotheses in spatial process models. On the one hand, this is necessary for modified Likelihood Ratio procedures, such as the Cox test. However, in spatial analysis, explicit nonlinear ML is also necessary for tests based on artificial nesting, this in contrast to the situation in a time series context. As pointed out in Sections 6.1. and 6.2, this is due to the presence of spatially lagged dependent variables.

As shown in Anselin (1984a), the generalization of a Cox–type test to spatial autoregressive specifications is not straightforward. In Pesaran (1974) and Pesaran and Deaton (1978), the Cox test for a standard regression model is found as:

$$T = (N/2).\ln[(\sigma_j)^2/(\sigma_{ji})^2]$$

where the index ji refers to the second model (j), estimated assuming that the first one (i) is the correct specification. Due to the properties of least squares residuals, the variance $(\sigma_{ji})^2$ can be found as:

$$(\sigma_{ji})^2 = (\sigma_i)^2 + (\sigma_a)^2$$

where $(\sigma_i)^2$ is the estimated residual variance in model i, and $(\sigma_a)^2$ is the estimated residual variance in a regression of the predicted values of the first model, $X_i b_i$, on the explanatory variables of the second, X_j.

In models with spatially lagged dependent variables, these simplifying results no longer hold, and a full set of asymptotic conditions needs to be solved. The resulting statistic is:

$$T_i = (N/2).\ln[(\sigma_j)^2/(\sigma_{ji})^2] + \ln \{|I-\rho_{ji}W_j|/|I-\rho_j W_j|\}$$

where the subscripts have the same meaning as before. In order to implement this statistic, an estimate for its variance is needed. In contrast to the results for time series models, and for regressions with serial autocorrelation considered in Walker (1967) and Pesaran (1974), the derivation of this variance in the presence of spatial dependence is highly nonlinear, and cannot be expressed in simple analytic terms. Consequently, this approach is not practical for spatial models. [11]

Asymptotically equivalent procedures, which turn out to be computationally much simpler, can be based on the artificial nesting approach. This boils down to a test on the significance of a nesting coefficient α, in an augmented model. Formally:

$$y = (1-\alpha).f_i + \alpha.g_j + \epsilon$$

where f_i is the specification under the null, and g_j is the predicted value for the alternative specification f_j, with its parameters replaced by their ML estimates.

The predicted value g_j can be obtained in the usual fashion, from a regression of y on f_i, as in the J−test of Davidson and MacKinnon (1981). Alternatively, g_j can be based on the regression of the second model f_j under the null hypothesis, as originally suggested by Atkinson (1970). This is approximated by the use of the predicted values in a regression of g_i on f_j, as in the JA−test of Fisher and McAleer (1979).

The significance of the nesting parameter can be assessed by means of the usual asymptotic Wald or Likelihood Ratio tests, or can be based on the exact results of McAleer (1983).

In a nonspatial situation, the equivalence of least squares estimation and maximum likelihood can be exploited to carry out these tests as straightforward extensions of most commonly used regression packages. In models with spatially lagged dependent variables, this is no longer the case. At first sight, it would seem that the asymptotic properties of the J and JA tests may not hold for spatial models. However, as shown in Anselin (1986a), the application of results by MacKinnon, White and Davidson (1983) allows for the inclusion of lagged dependent variables, provided that the generating process satisfies the proper mixing conditions. In general terms, and translated to the spatial context, these conditions imply a bounded variance, and a spatial dependence which decreases as the distance between observations increases.   These conditions are satisfied by most specifications of interest, as pointed out in Section 5.1.4.

In Anselin (1984b), the J and JA test are applied to a simple spatial autoregressive function of aggregate housing values. Four different forms for the spatial dependence are considered, reflected in a first−order contiguity (W_1), second order contiguity (W_2), a full inverse distance (W_3), and an inverse distance with a cut−off of two miles (W_4). The models are estimated for the same 49 contiguous neighborhoods in Columbus, Ohio, that are illustrated in Figure 12.1. The results of maximum likelihood estimation are listed in Table 13.1. The autoregressive coefficients have similar values for models 1, 3, and 4, but varying significance, with only the results for 1 and 4 being clearly significant. The fit of the four models, as measured by the maximized likelihood is very similar, and does not provide a clear distinction to eliminate one of the four specifications as unsatisfactory.

Table 13.1.   Estimation of a Spatial Autoregressive Model of Housing Values (asymptotic t−values in parentheses).

$$y = 24.178 +0.372 \ W_1 y \qquad\qquad L = -209.7$$
$$(3.643) \ (2.325)$$

$$y = 38.598 -0.004 \ W_2 y \qquad\qquad L = -211.9$$
$$(4.027) \ (0.019)$$

$$y = 23.752 +0.386 \ W_3 y \qquad\qquad L = -211.7$$
$$(1.903) \ (1.224)$$

$$y = 23.155 +0.411 \ W_4 y \qquad\qquad L = -210.1$$
$$(2.934) \ (2.153)$$

Source: Anselin (1984b), p. 101.

The non−nested tests provide a means to evaluate each of the structures in more detail, and to attempt to falsify each by using the others as an alternative. The test values, which are asymptotically distributed as standard normal, are listed in Table 13.2. In each row, a significant test statistic points to the falsification of the model in the row by the model in the column.

As the results in Table 13.2. show, the asymptotically equivalent tests do not yield the same significance in finite samples. Specifically, the JA test tends to lead to more rejections than the J−test. Also, the indication given by the statistics often implies symmetry, i.e., that each model in turn is able to falsify the other. Potential problems of interpretation associated with this are discussed further in Section 13.4.4. [12]

Table 13.2. ML–Based Non–Nested Tests of Model i (Row)
Against Model j (Column).

		Model 1	Model 2	Model 3	Model 4
1.	J		0.416	0.440	1.857
	JA		0.241	3.394	3.749
2.	J	2.992		1.561	2.847
	JA	−0.951		−2.366	−2.828
3.	J	3.072	0.151		2.898
	JA	4.195	2.852		5.338
4.	J	2.616	0.695	0.246	
	JA	4.299	1.699	4.323	

Source: Anselin (1984b), p. 102.

13.4.3. Non–Nested Tests on the Spatial Weight Matrix Based on Instrumental Variable Estimation

As shown in Chapter 7 and illustrated in Section 12.3.3, inference in models with spatially lagged dependent variables can also be based on instrumental variable or 2SLS estimation. This allows many additional tests on non–nested hypotheses to be applied to spatial process models. One category of such tests is based on the similarity of a likelihood to a quadratic form in IV residuals: [13]

$$(y - Xb)'Q(P)(y - Xb)$$

where b are IV estimates, obtained as:

$$b = [X'Q(P)X]^{-1}X'Q(P)y$$

with

$$Q(P) = P[P'P]^{-1}P'$$

as an idempotent projection matrix, and P as a matrix of instruments. Using this notation for the spatial process models would include the spatially lagged dependent variable among the columns of X.

Tests on non–nested hypotheses based on IV estimation were suggested by Godfrey (1983) and Ericsson (1983). Both authors point out that the difference of quadratic forms in the residuals forms a natural analogue to the modified Likelihood Ratio test of Cox. The general idea behind the Godfrey and Ericsson tests is the same. The differences between the tests follow from the use of alternative asymptotic approximations. All the resulting statistics are asymptotically standard normal.

In applying these tests to spatial models, no special adaptations are needed. The only operational concern is the choice of a proper set of instruments, which have to be the same for the model in H_i as well as in H_j.

The form proposed by Godfrey is:

$$G_i = N^{1/2}.[(D_i/N) - d_i]/(v_i)^{1/2}$$

with

$$D_i = e_j'Q(P)e_j - e_i'Q(P)e_i$$

as the difference of quadratic forms in the residuals of each model (e_i and e_j). As before, the subscripts refer to each model. To obtain the other expression in G_i, some additional notation is needed:

$$Z_{i(j)} = Q(P).X_{i(j)}$$

i.e., the Z are predicted values for each explanatory variable, obtained from a regression of each X on the instruments in P. Also,

$$M(Z) = I - Z[Z'Z]^{-1}.Z'$$

$$Z^*_{i(j)} = M(Z_j).Z_i$$

where $M(Z)$ is also idempotent, and Z^* consists of the residuals in a regression of each of the predicted instrument values in i (Z_i) on all the predicted instrument values of j (Z_j). Further auxiliary expressions are:

$$d_i = [b_i'Z_i'M(Z_j).Z_i b_i]/N$$

$$v_i = 4(\sigma_i)^2.[b_i'Z^*_i'.M(Z_i).Z^*_i.b_i]/N$$

and $(\sigma_i)^2$ as the ML estimate of the error variance.

The Ericsson tests are:

$$T_4 = c_4/(v_4)^{1/2}$$

$$T_6 = c_6/(v_4)^{1/2}$$

with as auxiliary expressions:

$$S_i = X_i[X_i'Q(P)X_i]^{-1}X_i'Q(P)$$

$$R_i = Q(P).S_i$$

$$(\sigma_i)^2 = e_i'e_i/(N-K_i)$$

$$c_0 = y'[Q(P) - R_i]y/[(\sigma_i)^2.N^{1/2}]$$

$$c_4 = -y'\{[Q(P) - R_j] - S_i'[Q(P) - R_j]S_i\}y/[(\sigma_i)^2.N^{1/2}]$$

$$c_6 = c_4 + c_0$$

$$v_4 = 4.b_i'X_i'S_j[Q(P) - R_j]S_jX_ib_i/[N.(\sigma_i)^2].$$

Alternative approaches are based on the artificial nesting principle. The J—test can be applied in the same manner as before, in an augmented regression which includes the predicted value from the second model, g_j. The only difference with the ML approach is that now estimation and significance tests are based on instrumental variables. A slightly different test is suggested by Godfrey (1983), and labeled T_X. It consists of a significance test on the artificial nesting parameter for g_j, where g_j is taken as:

$$g_j = Z_i^*.b_i$$

in the same notation as before.

The various IV—based tests are applied to some simple spatial models of aggregate housing values in Anselin (1984a). To illustrate the differences between these approaches, consider two specifications in particular, each containing a different structure for the spatially lagged dependent variable, as well as different explanatory variables. The first model explains aggregate housing values for planning neighborhoods by a spatial variable with first order contiguity. In addition, income (I), an indicator of crime (C) and of general housing quality (P, plumbing) are included as explanatory variables. The second specification includes inverse distance spatial dependence, income, open space (S) and plumbing. Estimation results for these simple models are listed in Table 13.3. The values for the various test statistics are given in Table 13.4. for the G_i, T_4 and T_6 tests, and in Table 13.5. for the J and T_X test. The latter shows both the estimate for the nesting parameter and the corresponding test value. [14]

Table 13.3. **Instrumental Variable Estimates for Models of Aggregate Housing Value (t—test values in parentheses)**

$$y = 29.333 \ +0.497W_1y \ +0.179\ I \ -0.533\ C \ +1.468\ P$$
$$(1.945) \ (1.727) \quad (1.649) \quad (3.283) \quad (2.678)$$
$$R^2 = 0.46$$

$$y = -10.742 \ +0.875W_2y \ +0.317\ I \ +0.747\ S \ +1.248\ P$$
$$(0.871) \ (2.946) \quad (2.708) \quad (1.412) \quad (1.741)$$
$$R^2 = 0.35$$

Source: Anselin (1984a), p.176.

Table 13.4. Nonnested Tests Based on an IV−Analogue to the Likelihood Ratio.

		Model 1	Model 2
1.	G_i		−6.001
	T_4^i		1.402
	T_6		5.687
2.	G_i	−8.880	
	T_4^i	1.406	
	T_6	8.415	

Source: Anselin (1984a), p. 177.

Table 13.5. Nonnested Tests Based on Instrumental Variable Estimation in an Augmented Regression.

		Model 1	Model 2
1.	a_J		0.865
	J		1.289
	a_T		−5.856
	T_X		1.000
2.	a_J	0.915	
	J	2.035	
	a_T	−9.294	
	T_X	−1.749	

Source: Anselin (1984a), p. 176.

Similar to the results reported in Table 13.2, the asymptotic equivalence of the various tests does not translate into comparable finite sample properties. The indications given by the five tests differ quantitatively but also qualitatively. For example, the second model is falsified by the first according to four out of the five tests. The reverse is indicated for only two of the tests. The potential conflict between the conclusions drawn from the different tests illustrates some of the problems which may be faced in the use of non−nested tests in spatial econometrics. I turn to this issue next.

13.4.4. Evaluation and Practical Implications

The insight gained from the use of the nonnested approach in specification tests for the structure of spatial dependence in the weight matrix is not always unambiguous. This is particularly the case when several alternative specifications are considered, and no clear a priori motivations exist to prefer one over the other. The result that two models are able to falsify each other, or that neither is able to falsify the other is not easy to interpret. On the one hand, this may be an indication of the weakness in both theoretical constructs, i.e., they are poor theories, easy to falsify. Alternatively, as illustrated by the results of several small sample simulations, it may indicate a lack of power of the tests, i.e., the tests cannot distinguish between different models.

When several models are tested in pairwise fashion, the interpretation can easily become complex. In principle, since the tests are not model discrimination procedures per se, the information provided by a set of test results should not be interpreted as the basis for selecting one model as best. Indeed, when several tests are used, they should be considered as multiple comparisons, and the appropriate adjustments to the critical significance levels should be carried out. However, in applied work, this is easily overlooked, and it is often more relevant to somehow use the test indications as a basis to select a model.

In a soft modeling approach, this can be achieved by summarizing the information in a qualitative pairwise comparison matrix of non−nested test results, as illustrated in Anselin and Can (1986). Obviously, the formal statistical foundations of the non−nested tests do not provide a rigorous basis for this, and it is largely an ad hoc procedure. However, in practice one may be willing to sacrifice some of the rigor in order to achieve greater qualitative insight. Hence, based on a soft but structured decision analytic approach, it is possible to summarize the different test results in terms of a concordance index for a pairwise comparison matrix, or a similar measure. This, in turn, can be incorporated into an overall subjective model validation context.

Another issue which may lead to problems in applied work concerns the asymptotic nature of most of the test statistics. As evidenced in several Monte Carlo simulation experiments reported in the literature, the asymptotic equivalence of the tests does not necessarily carry over in finite samples. Not only does the small sample power of the tests vary considerably, but they also often lead to conflicting indications about the validity of the models considered. Neither characteristic is very attractive in applied work. [15]

In addition, as pointed out in Anselin (1984a), many of the finite sample corrections suggested in the econometric literature do not carry over in a straightforward way to models with spatial dependence. While further research is needed on this issue, it is doubtful that the results would be very relevant in practice. Indeed, from an applied perspective, the gain in insight from using these rather sophisticated techniques for distinguishing among mostly rudimentary models may not be worth the cost.

Overall, the degree of sophistication and inherent limitations of the nonnested tests may seriously hamper their usefulness in applied work. Particularly, when the specifications of interest are largely ad hoc, or when the number of observations is small, the test results may be highly unreliable. In those instances, other criteria of model validity are to be preferred, such as the model selection techniques discussed in the next chapter.

NOTES ON CHAPTER 13

[1]   Chapters 13 and 14 are based in a large part on the ideas formulated in Anselin (1987a).

[2]   For example, in economics, issues such as the largely non—experimental nature of empirical analysis, the complexity of human behavior, and the paucity of adequate data have led to recurrent criticisms of the soundness of the methodological framework on which the estimation, testing and evaluation of econometric models is built. The literature on this issue is extensive, and goes back to early discussions between Keynes and Tinbergen (Keynes 1939; Tinbergen 1940, 1942). More recently, econometric practice, the performance of econometric models, and the use of ad hoc specifications and data mining have been critically evaluated in, e.g., Haitovsky and Treyz (1970, 1972), Dhrymes et al. (1972), Shapiro (1973), Leamer (1974, 1978, 1983), Christ (1975), Mayer (1975, 1980), Zellner (1979, 1984), Sims (1980), Kmenta and Ramsey (1980), Cooley and LeRoy (1981, 1985, 1986), Frisch (1981), Malinvaud (1981), Chow and Corsi (1982), Lovell (1983), Peach and Webb (1983), Ziemer (1984), and Swamy, Conway, and von zur Muehlen (1985).

[3]   For a more extensive discussion of the relevance of these issues to theory in regional science and geography, see, e.g., Isard (1969), Amadeo and Golledge (1975), Weibull (1976), Isard and Liossatos (1979), Golledge (1980), Cox and Golledge (1981), Paelinck, Ancot, and Kuipers (1982), Couclelis and Golledge (1983), and Beckmann and Puu (1985).

[4]   For recent overviews of these general issues, see, e.g., Ramsey (1974), Hausman (1978), Thompson (1978a, 1978b), Amemiya (1980), Bierens (1982), Engle (1984), Ruud (1984), Tauchen (1985), Newey (1985a, 1985b), and Davidson and MacKinnon (1985).

[5]   The direction from complex to simple model has the advantage that inference in overparameterized models, when the correct specification has a smaller number of coefficients, is not biased but only less efficient. On the other hand, inference in an underspecified model may be seriously biased.

[6]   The study of agricultural consumption in Irish counties by O'Sullivan, as originally reported and analyzed in Cliff and Ord (1981, pp. 208—210, and p. 230). See also Anselin (1988a) for additional insight into specification analysis for spatial models using this data set.

[7]   Burridge used an inverse distance form for W, while Blommestein applied higher order spatial lags. A series of specification tests based on the Lagrange Multiplier approach in Anselin (1988a) confirm the results obtained by Bivand, and further emphasized the importance of the choice of the weight matrix relative to the selection of the lag length.

[8]   For a more in—depth analysis of the multiple comparsion issue, see also Meeks and D'Agostino (1983).

[9]   This section is based extensively on the discussion in Anselin (1984a, 1984b, 1986a).

[10]   A technical discussion of the interpretation of non—nested tests, as compared to the more familiar nested approach is given in, e.g., Fisher and McAleer (1979), Dastoor (1981), Fisher (1983), Gourieroux, Monfort, Trognon (1983), MacKinnon (1983), McAleer and

Pesaran (1986), and McAleer (1987). An alternative viewpoint is presented in the recently developed encompassing principle of Mizon and Richard (1986), which includes non—nested tests as a special case.

[11] For further details, see Anselin (1984a).

[12] For more details on the empirical results, see Anselin (1984b).

[13] See Gallant and Jorgenson (1979) for a formal motivation for this approach.

[14] For a more detailed description of variable definitions, as well as additional results, see Anselin (1984a).

[15] For Monte Carlo results in standard econometrics, see, e.g., Pesaran (1982b), Davidson and MacKinnon (1982), and Godfrey and Pesaran (1983). In Anselin (1986a), a number of simulations are carried out for the ML—based J and JA tests in spatial autoregressive models. The results are in general agreement with those for the non—spatial case.

CHAPTER 14

MODEL SELECTION IN SPATIAL ECONOMETRIC MODELS

In this chapter, the statistical validation of spatial econometric models is approached from a different perspective, based on model selection or model discrimination techniques. In this context, the issue of selecting one model from among a number of alternative candidates is considered as a decision problem. More specifically, attention is focused on the nature of the trade—off between the number of parameters included in the model and the model fit, in the sense of how well the estimated model predicts in—sample observations.

The chapter consists of five sections. After a brief discussion of appropriate measures of goodness—of—fit for spatial models, I focus in more detail on information based criteria, Bayesian approaches and heuristics. These different perspectives on model selection can be applied to single equation specifications as well as in more general contexts. In contrast to the specification tests discussed in the previous chapter, they are not specifically geared to *spatial* applications.

These techniques have not yet found general acceptance in the practice of empirical regional science, although they are increasingly suggested in the literature as ways to avoid some of the pitfalls of traditional methods. As such, they also form a starting point for several promising and unexplored research directions in spatial econometrics.

The chapter is concluded with a brief discussion of some operational and practical concerns about model selection in spatial econometrics.

14.1. Measures of Fit in Spatial Models

The assessment of the empirical fit of an estimated model is an important aspect of econometric analysis. In spatial econometrics, this is slightly more complex due to the lack of a standard measure such as the R^2. Although this measure is routinely supplied in empirical work, its interpretation in the presence of spatial effects may be misleading, and it no longer has a direct link to a test on the overall significance of the estimated model (F—test). Two different situations can be distinguished, one where spatial dependence is present in the error term, the other where spatially lagged dependent variables are included in the model.

In the first case, the same characteristics are obtained as for a model with a general nonspherical error variance. Specifically, since estimation is based on EGLS or ML, the residuals of the estimated model may not have a zero mean, and the standard decomposition of observed variance into explained and residual variance no longer holds. An R^2 measure that is calculated in the usual way is therefore meaningless, and may yield nonsensical values. This is the case both for a mean—adjusted and for a non mean—adjusted R^2. In addition, the equivalence of R^2 to the squared correlation between the observed and predicted values becomes invalid. [1]

A number of pseudo R^2 values can be reported, in the form of alternatives that mimic certain aspects of the standard measure. For example, the squared correlation between the observed and predicted dependent variable still provides a measure of linear association (or in–sample predictive ability) that is between zero and one, although it no longer is related to the variance decomposition. Alternatively, a pseudo R^2 can be based on weighted predicted values and residuals, as in Buse (1973, 1979). [2] For the model with a spatial autoregressive error term, this R^2 becomes:

$$R^2 = 1 - e'(I-\lambda W)'(I-\lambda W)e \ / \ (y-\iota y_w)'(I-\lambda W)'(I-\lambda W)(y-\iota y_w)$$

where

$$y_w = \iota'(I-\lambda W)'(I-\lambda W)y \ / \ \iota'(I-\lambda W)'(I-\lambda W)\iota$$

with ι as an N by 1 vector of ones. When the weight matrix is row–standardized, the numerator in y_w becomes $N(\lambda-1)^2$, since $W\iota=\iota$.

When a spatially lagged dependent variable is included, estimation is based on maximum likelihood, or on an instrumental variables approach. In the latter case, the residuals have a zero mean, so that the standard variance decomposition can be obtained, and an R^2 can be computed in the usual manner. However, the link between this R^2 and an exact test of significance no longer exists, due to the asymptotic nature of inference in IV models. When estimation is based on ML, the standard R^2 is invalid. A more appropriate measure of fit is based on the maximized log–likelihood, although a pseudo R^2, in the form of a squared correlation between predicted and observed values can also be used.

As an illustration, the R^2, squared correlation (Corr), pseudo–R^2 ($P-R^2$) and maximized log–likelihood (L) for the spatial models estimated in Section 12.2. are presented in Table 14.1. The pseudo–R^2 is listed only for those specifications where it is applicable, i.e., in models with a non–spherical error term. For those models estimated by OLS, i.e., the simple regression and the spatial expansion, the R^2 and squared correlation are the same. Those are also the only specifications where the unadjusted R^2 is appropriate. For the other models, the R^2 is listed only to illustrate the extent to which its uncritical application may lead to misleading indications of fit.

Clearly, the ranking of model fit which follows from the various measures is far from uniform. For example, only one ranking is the same between the squared correlation and the log–likelihood. According to both criteria, the Spatial Durbin model is best. However, the log–likelihood shows the model with spatial dependence and heteroskedasticity as second, whereas it ranks last according to the squared correlation. There is slightly more agreement between the naive R^2 and the squared correlation, although, in all, only three out of seven models rank the same for both measures. The relative ranking indicated by the psuedo–R^2 is the same as for the squared correlation, but does not agree with the log–likelihood. In sum, considerable caution is needed when interpreting these various measures of fit as indicators of model validity.

Table 14.1. Measures of Fit for Spatial Models of Neighborhood Crime (Ranking in Parentheses)

Model	R^2	Rank	Corr.	Rank	$P-R^2$	Rank	L	Rank
1.	0.55	(5)	0.55	(4)			−170.40	(7)
2.	0.62	(3)	0.65	(2)			−165.41	(3)
3.	0.32	(6)	0.54	(6)	0.42	(2)	−166.40	(5)
4.	0.63	(2)	0.67	(1)			−164.41	(1)
5.	0.63	(1)	0.63	(3)			−165.51	(4)
6.	0.61	(4)	0.55	(4)	0.55	(1)	−169.77	(6)
7.	0.30	(7)	0.52	(7)	0.38	(3)	−164.69	(2)

Models: (1) Simple Regression, Section 12.2.2.; (2) Mixed Regression Spatial Autoregression, Section 12.2.3.; (3) Spatial Dependent Errors, Section 12.2.4.; (4) Spatial Durbin, Section 12.2.5.; (5) Spatial Expansion, Section 12.2.6.; (6) Heteroskedastic Errors, Section 12.2.7.; (7) Joint Spatial Dependent and Heteroskedastic Errors, Section 12.2.9.

Other traditional measures of goodness−of−fit, such as the chi−squared, MSE (mean squared error), and MAPE (mean absolute percentage error), have been applied to spatial models. Although they are computed in the usual fashion, their interpretation in a spatial context is not always without problems, e.g., as overviewed in Smith and Hutchinson (1981), Fotheringham and Knudsen (1986), and Knudsen and Fotheringham (1986).

From an econometric point of view, one of the problems with these measures is their lack of adjustment for the inclusion of additional explanatory variables, which tends to inflate the indication of fit. If used indiscriminately, such measures can provide misleading guidance, as pointed out in, e.g., Mayer (1975, 1980), Thompson (1978a, 1978b), Amemiya (1980), and Kinal and Lahiri (1984).

Several choice criteria have been suggested that take the trade−off between fit and parsimony into account in an explicit manner. These criteria can be distinguished by the objective function used as the basis for the choice, i.e., by the associated loss or risk function. Well known examples are the adjusted R^2, Mallow's C_p, S_p, and other prediction criteria, which are routinely provided by econometric computer packages. [3] However, these adjusted measures need to be interpreted with caution, since they may provide misleading indications in the presence of spatial dependence, as illustrated for Mallow's C_p in Anselin and Griffith (1988).

14.2. Information Based Criteria for Model Discrimination

Information theoretic measures are directly related to the concept of entropy, which underlies a substantial body of work on spatial interaction theory and operational urban and regional models. The use of information theory as a tool in model building and model validation in regional science and geography has been advocated from a number of different viewpoints, e.g., in Wilson (1970), Sheppard (1976), Snickars and Weibull (1977), Buck and Hakim (1980), Thomas (1981), and Ayeni (1982, 1983). In the econometric literature, information theory is used to obtain a measure of closeness of the assumed model to the unknown *true* model.

Operationally, competing models under consideration are conceptualized as probability density functions in the observed dependent variable. A distinction is made between the density for the true but unknown model g(y), and the postulated model h(y|X,θ). As usual, the latter is conditional upon the explanatory variables X and the parameters θ.

An information–theoretic measure for goodness–of–fit is the Kullback–Leibler information criterion or KLIC. The KLIC is similar to an expected logarithmic likelihood ratio of the true model to the postulated model, evaluated with respect to the unknown true underlying distribution G:

$$KLIC = E_G \{\log [g(y)/h(y|X,\theta)]\}.$$

As argued by Akaike (1974, 1981) and Sawa (1978), the negative KLIC, which is equivalent to the entropy of the postulated model with respect to the true model, forms a natural measure of fit. Since the expected value in the KLIC, or entropy, involves an unknown distribution, a practical criterion needs to be based on an appropriate estimate. The Akaike Information Criterion (AIC) and related measures such as BIC are estimates for this expected value which maximize entropy or predictive likelihood. [4]

In general terms, this measure can be formally expressed as:

$$AIC = - 2.L + q(K)$$

where L is the maximized logarithmic likelihood, K is the number of unknown parameters of the model, and q is a correction factor. The correction factor varies among the different versions of the information measure. In the original formulation of Akaike, a simple multiplicative factor with q=2K is used. In the versions advanced by Hannan and Quinn, and Schwarz, q=logN.K (with N as the number of observations) and q=2.loglogN.K. The general idea is always that the AIC corrects or penalizes the assessment of goodness–of–fit given by the maximized likelihood by a factor which reflects the number of parameters.

The AIC has a certain similarity to the Bayesian notions of posterior predictive probability and posterior odds, as pointed out by, e.g., Zellner (1978), Leamer (1979), Atkinson (1978, 1980, 1981), and Chow (1981). Although the resulting expressions are essentially of the same form, they involve basically differing viewpoints. [5] The AIC is also often compared to the C_p statistic of Mallows (1973), which is based on a different prediction criterion, i.e., the mean squared error of prediction. [6]

The AIC and its related measures have recently been used in a number of empirical analyses in regional science: studies of criminal behavior in Buck and Hakim (1980, 1981), spatial interaction models in Anselin (1984c), spatial autoregressive models in Bivand (1984), and switching regression models of urban densities in Brueckner (1985, 1986). The implementation of information based criteria of model validity in spatial analysis is straightforward, provided that the likelihood function properly incorporates the relevant spatial effects (e.g., spatial dependence). This is illustrated in Table 14.2, where for four of the specifications from Section 12.2, the log–likelihood and corresponding AIC and SC (Schwartz Criterion) are listed. The four models are the simple regression, and three ways of incorporating spatial dependence: a mixed regressive spatial autoregressive model, a specification with spatially dependent errors, and a Spatial Durbin model.

The ranking of models implied by both information measures of fit is the same, although the relative magnitudes for the AIC and SC are clearly different. For example, the difference between the second and third model (spatial dependent errors and Spatial Durbin) is much smaller according to the AIC measure than following the SC, illustrating the effect of the use of a different loss function. Also, the trade–off between fit and parsimony, which is encompassed in the AIC and SC, yields a different indication of model validity than the simple log–likelihood. According to the latter, as in Table 14.1, the Spatial Durbin model is best. However, since the AIC and SC take into account the relative lack of additional fit provided by the two extra explanatory variables in this model, its corresponding ranking is affected, and it is shown to be worse than the other two spatial models. Overall, the mixed regressive spatially autoregressive model achieves the highest validity.

Table 14.2.   Information Based Measures of Fit for Spatial Models of Neighborhood Crime (Ranking in Parentheses)

Model	L	Rank	AIC	Rank	SC	Rank
1.	−170.40	(4)	346.79	(4)	352.47	(4)
2.	−165.41	(2)	338.82	(1)	346.38	(1)
3.	−166.40	(3)	340.80	(2)	348.36	(2)
4.	−164.41	(1)	340.82	(3)	352.17	(3)

Models: (1) Simple Regression, Section 12.2.2.; (2) Mixed Regression Spatial Autoregression, Section 12.2.3.; (3) Spatial Dependent Errors, Section 12.2.4.; (4) Spatial Durbin, Section 12.2.5.

A few issues may complicate the interpretation of the information based measures in applied work. Since the nature of the trade–off between parsimony and fit, i.e., the q in the expression for AIC, is based on asymptotic considerations and a specific loss function, it may not be clear which measure should be used in a particular situation. Moreover, the implied loss functions penalize differently for the inclusion of additional variables (different q) and could yield conflicting indications. In other words, different versions of the AIC correspond to using different ad hoc objective functions. To some extent, one could construct an adjusted measure in such a way that a particular model would be chosen as best. Consequently, caution is needed when interpreting the results. Ideally, the loss function should correspond to the objectives of the analysis or analyst, but this is not always implementable in practice. In an empirical situation, it is therefore preferable to calculate several measures and to assess the sensitivity of the results from the degree of conflict. On the theoretical side, the appropriateness of particular loss functions in spatial analysis needs to be addressed, e.g., to incorporate the importance of the spatial pattern of fit.

The AIC and related measures depend on the correct choice of the likelihood function and the results of maximum likelihood estimation. They may therefore not be very reliable in situations where the distributional assumptions are more out of convenience than based on solid theoretical considerations. In those instances robust measures are more appropriate.

14.3. Bayesian Approaches to Model Selection

As discussed in Section 7.2, in a Bayesian approach to model validation the data analysis as such is combined with the decision process inherent in the construction of hypotheses and choice between models. Formally, the evaluation of the validity of a model and the comparison of competing specifications is based on the use of posterior probabilities and odds ratios, obtained from a combination of the information contained in the data and the prior convictions of the analyst. [7]

Two competing models (hypotheses), represented as M_i and M_j, are related to observations on the dependent variable by means of a probability density function. The important concepts in a Bayesian terminology are then: the prior (subjective) probability for each model to be true —for M_i: $P(M_i)$ — the conditional probability of the data (likelihood), given that the model is correct, $P(y|M_i)$, and the posterior probability that the model is correct, conditional on the observed data, $P(M_i|y)$. Using Bayes' Law, the ratio of the posterior probabilities or the posterior odds ratio is obtained as the product of the prior odds ratio and the well known Bayes factor:

$$P(M_i|y)/P(M_j|y) = [P(M_i)/P(M_j)].[P(y|M_i)/P(y|M_j)].$$

When the models are expressed in function of parameters, the Bayes factor corresponds to a ratio of averaged integrals of likelihoods, where the prior distributions for the parameters (conditional on the model specification) are used as weights. To incorporate the posterior odds and probabilities in the decision framework, a loss function or utility function is specified, which expresses the consequences of the model selection (or parameter inference) for the analyst. The model selection itself is then based on maximizing the expected utility, or equivalently, on minimizing the expected loss.

The resulting framework provides great flexibility and is suitable in a variety of model validation contexts, such as the comparison of nested as well as non—nested hypotheses, and the choice between different underlying probability distributions. It also forms a consistent way to consider varying degrees of prior beliefs in the strength of the models under scrutiny.

A related concept is the idea of fragility analysis, developed by Leamer and Leonard (1983), which provides a way to assess the sensitivity of model estimates to alternative specifications. This idea may have particular relevance to the issue of the choice of a weight matrix for spatial dependence. Compared to the current largely ad hoc approach to this problem, it may yield a superior framework. Although some initial work has been started in this direction, substantial further research is needed.

The implementation of a Bayesian approach in applied regional science is not without problems. A main concern pertains to the determination of prior probabilities, and the extent to which these should incorporate information other than that contained in the data. There are methodological problems associated with the derivation of the proper diffuse or non—diffuse priors with the desired characteristics, as pointed out in Zellner (1971, 1984), Kadane et al. (1980), and Klein and Brown (1984). In addition, as evidenced by a large literature in psychometrics and decision analysis, the operational assessment of the analyst's subjective probabilities is complex. [8] Furthermore, the expected utility framework on which the decision analysis is based, is not without its drawbacks, and is not necessarily representative of actual human decision behavior. [9] Also, as pointed out in Section 7.2, the implementation of Bayesian techniques in instances with a large number of variables, parameters and models is limited by its reliance on numerical integration, and the constraints imposed by computational technology.

14.4. Heuristics in Spatial Model Selection

A growing awareness of the frequent failure of the standard assumptions of parametric models in applied empirical work, e.g., due to the uncertainty about the correct underlying distribution, measurement problems and poor data quality, has led to a greater emphasis on non—parametric, soft techniques and heuristic approaches to data analysis. Some robust approaches to the estimation of spatial process models were illustrated in Section 7.3. In this section two approaches to model validation are considered.

One approach, originally suggested by Stone (1974) and Geisser (1975) is based on resampling techniques, similar to the jackknife and bootstrap discussed in Section 7.3. The idea behind this Stone—Geisser cross—validation strategy is simple, and consists of assessing the validity of a set of competing models by comparing their predictive ability on data which have not been used in the model estimation. More specifically, by removing observations or sets of observations one at a time and estimating the competing models on the reduced data set, a specification is selected which best predicts the complete data set. Provided that spatial dependence among observations is properly accounted for, this procedure could be applied in a wide range of spatial econometric specification analyses and provide a robust basis for model selection.

Another approach deals with the comparison of related data matrices, based on work in the psychometric literature on the properties of dependent cross product correlation coefficients. It was developed by Hubert and Golledge (1981, 1982) and applied to a wide range of spatial models. As a special case of the quadratic assignment problem, the approach results in a non–parametric measure of fit for a variety of combinations of theoretical (predicted) and actual data matrices. Formally, the coefficient used to compare matrices A, B and C is expressed as:

$$\rho_{A,B-C} = (\rho_{AB} - \rho_{AC}) \ / \ [2(1 - \rho_{BC})]^{-1/2}$$

where ρ is a cross–product correlation coefficient or other index of correspondence.
10

The reference distribution of the statistic is formed from a randomization strategy, by simultaneously permuting matrix rows and columns. As a result, the statistic can provide an insight into the relative fit to a given data set of a number of competing models, and is not affected by any distributional assumptions.

This general methodology has been used for the comparison of migration models in Gale et. al. (1983) and of proximity matrices of crime data in Golledge, Hubert and Richardson (1981).

The resulting measure of fit is extremely flexible, easily implementable in practice and does not suffer from potentially erroneous assumptions about the underlying distribution   On the other hand, the fact that the validity of the reference distribution obtained by the randomization is limited to the given data set, may restrict the generality of the statistic.   Also, as the choice of an index of correspondence is to some extent arbitrary, it may be possible to obtain conflicting results from using different indices.   Nevertheless, this technique provides a powerful alternative in situations where the parametric assumptions may not be reliable.

14.5. Practical Implications of Model Validation in Spatial Econometrics

The three categories of model selection techniques discussed in the previous sections represent clearly distinct philosophies about data analysis. In a strict sense, since they are based on different frames of reference, the various indices of fit are not fully comparable. In applied work, this should be kept in mind, even though useful insights can be obtained from calculating more than one indicator of model validity.

Of the three, the Bayesian approach is at this point the least practical for spatial analysis, and also the hardest to implement. In part this is due to the analytic complexities and numerical difficulties discussed before. In addition, for it to be more useful for empirical regional science, several methodological issues need to be resolved, for example, to develop useful priors that take into account the multidirectional dependence in spatial analysis. On the other hand, the Bayesian viewpoint forms the most consistent parametric framework for carrying out specification searches, and for combining data analysis with the decision process of the analyst. It therefore deserves more research attention in spatial econometrics than it has received so far.

The information criteria and the heuristics are easy to implement in practice. While the former may not be reliable when the likelihood function is not specified properly, e.g., when the spatial dependence in the data is not acknowledged, the latter are robust. The matrix comparison procedures of Hubert and Golledge in particular, form a very attractive alternative in situations where the parametric assumptions may be suspect, and should be implemented routinely in an exploratory context. However, as pointed out in Section 7.3, most of the other robust techniques have been developed for situations of independent observations, and need further adaptation to take into account the multidirectional dependence in spatial analysis. Since the robustness is a feature which is more appropriate in the often ad hoc modeling environment in regional science, the extension of these techniques to deal with the spatial aspects of data analysis is an important research direction.

The decision problem inherent in the choice of the proper frame of reference is seldom taken into account in applied spatial econometrics. Typically, one approach is implemented and the others are ignored. However, it should be clear that the indications provided by the alternative frameworks could provide useful insights, and therefore should not be readily discarded. Their incorporation in the overall decision process of the analyst probably requires the explicit formulation of a multicriteria, potentially qualitative framework. The implementation of such a framework in spatial econometrics remains largely unexplored and an important area of future research.

As a summary of the various techniques, and to put the material in an applied perspective, I conclude this chapter with the following seven general guidelines for model validation in spatial econometrics:

[1] Mechanistic interpretation and uncritical use of standard output from econometric computer packages should be avoided. Space does make a difference, and several standard techniques need to be adjusted to take this into account.

[2] Results should not be given more weight than merited by the particular frame of reference used. In particular, the effect of sequential tests and multiple comparisons on the significance levels of various statistics should be made explicit and fully reported.

[3] The assumptions underlying the methods and models used should be checked before the analysis is carried out, for example, by performing some exploratory analyses and diagnostic checks for normality and by measuring spatial correlation for different W matrices.

[4] When in doubt about the assumptions, robust procedures should be preferred.

[5] Alternative ways of assessing model validity should be implemented, and the degree of conflict or agreement between the results reported. This provides information to the reader or user of the models to carry out his or her own interpretation.

[6] Asymptotic results should be used with care, and small sample adjustments implemented when possible.

[7] Implicit assumptions about spatial structure (e.g., the W matrix in tests for spatial autocorrelation) should be reported and the sensitivity of the results assessed.

NOTES ON CHAPTER 14

[1] For a more extensive discussion of this familiar result, see, e.g. Judge et al (1985), pp. 29–32.

[2] This approach is also appropriate for the SUR models in Section 10.1.

[3] See Amemiya (1980), for a more extensive overview.

[4] For a more technical discussion, see, e.g., Sawa (1978), Schwarz (1978), Hannan and Quinn (1979), and Amemiya (1980).

[5] See also the discussion in Section 14.3.

[6] For a technical discussion see, e.g., Amemiya (1980), Atkinson (1981), and Kinal and Lahiri (1984).

[7] For an overview, see, e.g., Dreze (1972), Gaver and Geisel (1974), Geisel (1975), Leamer (1978), Geisser and Eddy (1979), Box (1980), Kadane and Dickey (1980), and particularly Zellner (1971, 1978, 1979, 1984, 1985)

[8] For an overview, see Wallsten and Budescu (1983).

[9] This is exemplified by the existence of several paradoxes, e.g. surveyed in Schoemaker (1982).

[10] For details, see Hubert and Golledge (1981, 1982), and Hubert (1984).

CHAPTER 15

CONCLUSIONS

For a book such as this one, it is hard to formulate a single conclusion that accurately captures the range of materials covered and concisely reflects the main results. Instead, it may be more useful to point out the distinguishing characteristics of this collection of methods and models, and to summarize the most relevant contributions.

A major objective of the book was to demonstrate that many spatial effects can effectively be approached from an econometric perspective. In line with this goal, an important underlying theme was the characterization of spatial effects in the regression error term as a special case of a general parameterized nonspherical disturbance covariance matrix. This led to the development of several new tests and estimators for models that incorporate different combinations of spatial dependence and spatial heterogeneity, in cross—sectional as well as space—time data sets.

Another principal unifying concept was the application of the Lagrange Multiplier principle to develop several new tests for spatial effects. This approach, which only necessitates estimation under the null hypothesis of no spatial effects, led to tests that can be incorporated in standard regression packages with only minor additional computations. This greatly facilitates their implementation in applied regional analysis. In addition, the distributional properties of the tests are based on rigorous asymptotic foundations, instead of the rather ad hoc rationale underlying some traditional approaches. This Lagrange Multiplier approach was also applied to hitherto ignored situations of multiple spatial effects, such as the joint presence of spatial dependence and spatial heterogeneity.

In addition to the usual maximum likelihood approach, alternative techniques were covered as well, such as instrumental variables estimators, Bayesian methods, and various robust approaches.

Another overall theme underlying the book was the importance of illustrating how spatial effects influence and often invalidate the indications provided by standard econometric methods. In addition, the necessity of a realistic perspective on the validity of empirical spatial models is stressed. The book forms a general attempt to increase the awareness of the limitations of the current set of methodological tools. Although much progress has been made, substantial developments are still needed before spatial econometrics can take its place as a standard element among the methods of regional analysis.

I see some particularly promising directions for future research in the following six main areas:

[1]   An assessment of the relative merits of alternative paradigms for the analysis of spatial data, such as parametric analysis vs. nonparametrics, classical probability vs. Bayesian approaches, and exploratory vs. explanatory analysis.

[2]   A further evaluation of the appropriateness of an asymptotic framework for the analysis of realistic spatial data sets, and the development of practical finite sample approximations.

[3]   A more extensive exploration of alternative estimation and testing approaches. Particularly promising in this respect are robust techniques, such as the bootstrap, as well as Bayesian methods. In addition, the development of new techniques should be pursued; techniques that effectively take into account the pre−testing aspects of spatial analysis or that circumvent the need for using a spatial weight matrix.

[4]   A development of more effective diagnostics for multiple sources of misspecification in regression analysis, in the presence of spatial dependence and spatial heterogeneity.

[5]   An evaluation of the relative merits in realistic data contexts of the spatial econometric perspective and the spatial series approach. This could lead to a more fruitful integration of the two methodologies, similar to the development of ARMAX models and vector autoregressive specifications in time series analysis.

[6]   The development and dissemination of user−friendly software, which in many respects constitutes a necessary precondition for the effective diffusion of spatial econometric methods to the practice of regional science.

I hope that this book, by approaching a wide range of spatial aspects of empirical regional analysis from an econometric perspective, has made a contribution to the development of regional methods, useful for both applied econometricians as well as spatial analysts, and in line with the interdisciplinary tradition of regional science.

REFERENCES

Aguirre—Torres, V. and A. Gallant. (1983). "The Null and Non—Null Asymptotic Distribution of the Cox Test for Multivariate Nonlinear Regression: Alternatives and a New Distribution—Free Cox Test." *Journal of Econometrics*, 21, 5—33.

Akaike, H. (1974). "A New Look at Statistical Model Identification." *IEEE Transactions on Automatic Control*, AC 19, 716—23.

Akaike, H. (1981). "Likelihood of a Model and Information Criteria." *Journal of Econometrics*, 16, 3—14.

Alonso, W. (1968). "Predicting Best with Imperfect Data." *Journal of the American Institute of Planners*, 34, 248—55.

Amadeo, D., and R. Golledge. (1975). *An Introduction to Scientific Reasoning in Geography*. New York: Wiley.

Amemiya, T. (1980). "Selection of Regressors." *International Economic Review*, 21, 331—54.

Amemiya, T. (1985). *Advanced Econometrics*. Cambridge, MA: Harvard University Press.

Amemiya, T., and T. MaCurdy. (1986). "Instrumental—Variable Estimation of an Error—Components Model." *Econometrica*, 54, 869—80.

Amrhein, C., J. Guevara, and D. Griffith. (1983). "The Effects of Random Thiessen Structure and Random Processes on the Measurement of Spatial Autocorrelation." *Modeling and Simulation*, 14, 585—589.

Anderson, T., and C. Hsiao. (1981). "Estimation of Dynamic Models with Error Components." *Journal of the American Statistical Association*, 76, 598—606.

Anderson, T., and C. Hsiao. (1982). "Formulation and Estimation of Dynamic Models Using Panel Data." *Journal of Econometrics*, 18, 47—82.

Andrews, D. (1986). "A Note on the Unbiasedness of Feasible GLS, Quasi—Maximum Likelihood, Robust, Adaptive, and Spectral Estimators of the Linear Model." *Econometrica*, 54, 687—98.

Anselin, L. (1980). *Estimation Methods for Spatial Autoregressive Structures*. Ithaca, NY: Cornell University, Regional Science Dissertation and Monograph Series #8.

Anselin, L. (1981). "Small Sample Properties of Estimators for the Linear Model with a Spatial Autoregressive Structure in the Disturbance." *Modeling and Simulation*, 12, 899—904.

Anselin, L. (1982). "A Note on Small Sample Properties of Estimators in a First—Order Spatial Autoregressive Model." *Environment and Planning A*, 14, 1023—30.

Anselin, L. (1984a). "Specification Tests on the Structure of Interaction in Spatial Econometric Models." *Papers, Regional Science Association*, 54, 165—82.

Anselin, L. (1984b). "Specification of the Weight Structure in Spatial Autoregressive Models, Some Further Results." *Modeling and Simulation*, 15, 99—103.

Anselin, L. (1984c). "Specification Tests and Model Selection for Aggregate Spatial Interaction, an Empirical Comparison." *Journal of Regional Science*, 24, 1—15.

Anselin, L. (1985). *Specification Tests and Model Selection for Spatial Interaction and the Structure of Spatial Dependence*. Final Report to the National Science Foundation. Columbus, OH: Ohio State University Research Foundation.

Anselin, L. (1986a). "Non—Nested Tests on the Weight Structure in Spatial Autoregressive Models: Some Monte Carlo Results." *Journal of Regional Science*, 26, 267—84.

Anselin, L. (1986b). "Some Further Notes on Spatial Models and Regional Science." *Journal of Regional Science*, 26, 799—802.

Anselin, L. (1986c). "MicroQAP: A Microcomputer Implementation of Generalized Measures of Spatial Association." Discussion Paper, Department of Geography, University of California, Santa Barbara.

Anselin, L. (1987a). "Model Validation in Spatial Econometrics: A Review and Evaluation of Alternative Approaches." *International Regional Science Review*, 11, (forthcoming).

Anselin, L. (1987b). "Spatial Dependence and Spatial Heterogeneity, a Closer Look at Alternative Modeling Approaches." Working Paper, Department of Geography, University of California, Santa Barbara.

Anselin, L. (1987c). "Estimation and Model Validation of Spatial Econometric Models Using the GAUSS Microcomputer Statistical Software." Discussion Paper, Department of Geography, University of California, Santa Barbara.

Anselin, L. (1988a). "Lagrange Multiplier Test Diagnostics for Spatial Dependence and Spatial Heterogeneity." *Geographical Analysis*, 20, 1—17.

Anselin, L. (1988b). "Robust Approaches in Spatial Econometrics." Paper Presented at the Annual Meeting of the Association of American Geographers, Phoenix, AZ.

Anselin, L. and A. Can. (1986). "Model Comparison and Model Validation Issues in Empirical Work on Urban Density Functions." *Geographical Analysis*, 18, 179—197.

Anselin, L. and D. Griffith. (1988). "Do Spatial Effects Really Matter in Regression Analysis?" *Papers, Regional Science Association*, 65 (forthcoming).

Arora, S. and M. Brown (1977). "Alternative Approaches to Spatial Autocorrelation: An Improvement over Current Practice." *International Regional Science Review*, 2, 67—78.

Atkinson, A. (1970). "A Method for Discriminating Between Models." *Journal of the Royal Statistical Society B*, 32, 323—45.

Atkinson, A. (1978). "Posterior Probabilities for Choosing a Regression Model." *Biometrika*, 66, 39—48.

Atkinson, A. (1980). "A Note on the Generalized Information Criterion for Choice of a Model." *Biometrika*, 67, 413—8.

Atkinson, A. (1981). "Likelihood Ratios, Posterior Odds and Information Criteria." *Journal of Econometrics*, 16, 15—20.

Avery, R (1977). "Error Components and Seemingly Unrelated Regressions." *Econometrica*, 45, 199—209.

Ayeni, B. (1982). "The Testing of Hypothesis on Interaction Data Matrices." *Geographical Analysis*, 14, 79—84.

Ayeni, B. (1983). "Algorithm 11: Information Statistics for Comparing Predicted and Observed Trip Matrices." *Environment and Planning A*, 15, 1259—66.

Bahrenberg, G., M. Fischer, and P. Nijkamp. (1984). *Recent Developments in Spatial Data Analysis: Methodology, Measurement, Models*. Aldershot: Gower.

Balestra, P., and M. Nerlove. (1966). "Pooling Cross Section and Time Series Data in the Estimation of a Dynamic Model: The Demand for Natural Gas." *Econometrica*, 34, 585—612.

Baltagi, B. (1980). "On Seemingly Unrelated Regressions with Error Components." *Econometrica*, 48, 1547—51.

Baltagi, B., and J. Griffin. (1984). "Short and Long Run Effects in Pooled Models." *International Economic Review*, 25, 631—45.

Bar—Shalom, Y. (1971). "On the Asymptotic Properties for the Maximum Likelihood Estimate Obtained from Dependent Observations." *Journal of the Royal Statistical Society B*, 33, 72—77.

Bartels, C. (1979). "Operational Statistical Methods for Analysing Spatial Data." In *Exploratory and Explanatory Analysis of Spatial Data*, edited by C. Bartels and R. Ketellapper, pp. 5—50. Boston: Martinus Nijhoff.

Bartels, C. and R. Ketellapper (eds.). (1979). *Exploratory and Explanatory Analysis of Spatial Data*. Boston: Martinus Nijhoff.

Bartlett, M. (1975). *The Statistical Analysis of Spatial Pattern*. London: Chapman and Hall.

Bartlett, M. (1978). *An Introduction to Stochastic Processes, with Special Reference to Methods and Applications*. London: Cambridge University Press.

Bates, C., and H. White. (1985). "A Unified Theory of Consistent Estimation for Parametric Models." *Econometric Theory*, 1, 151–78.

Baxter, M. (1985). "Quasi–Likelihood Estimation and Diagnostic Statistics for Spatial Interaction Models." *Environment and Planning A*, 17, 1627–35.

Beckmann, M., and T. Puu. (1985). *Spatial Economics: Density, Potential, and Flow*. Amsterdam: North Holland.

Belsley, D., and E. Kuh. (1973). "Time–Varying Parameter Structures: An Overview." *Annals of Economic and Social Measurement*, 2/4, 375–8.

Belsley, D., E. Kuh, and R. Welsch. (1980). *Regression Diagnostics: Identifying Influential Data and Sources of Colinearity*. New York: Wiley.

Bennett, R. (1979). *Spatial Time Series*. London: Pion.

Bennett, R., R. Haining and A. Wilson. (1985). "Spatial Structure, Spatial Interaction and Their Integration. A Review of Alternative Models." *Environment and Planning A*, 17, 625–45.

Bera, A., and C. Jarque. (1982). "Model Specification Tests, a Simultaneous Approach." *Journal of Econometrics*, 20, 59–82.

Berndt, E., and N. Savin. (1977). "Conflict Among Criteria for Testing Hypotheses in the Multivariate Linear Regression Model." *Econometrica*, 45, 1263–72.

Berry, B. (1971). "Problems of Data Organization and Analytical Methods in Geography." *Journal of the American Statistical Association*, 66, 510–23.

Besag, J. (1974). "Spatial Interaction and the Statistical Analysis of Lattice Systems." *Journal of the Royal Statistical Society B*, 36, 192–236.

Besag, J. (1975). "Statistical Analysis of Non–Lattice Data." *The Statistician*, 24, 179–96.

Besag, J. (1977). "Efficiency of Pseudo–Likelihood Estimators for Simple Gaussian Fields." *Biometrika*, 64, 616–618.

Besag, J., and P. Moran. (1975). "On the Estimation and Testing of Spatial Interaction in Gaussian Lattice Processes." *Biometrika*, 62, 555–62.

Bhargava, A., and J. Sargan. (1983). "Estimating Dynamic Random Effects Models from Panel Data Covering Short Time Periods." *Econometrica*, 51, 1635–59.

Bhat, B. (1974). "On the Method of Maximum Likelihood for Dependent Observations." *Journal of the Royal Statistical Society B*, 36, 48–53.

Bickel, P. and D. Freedman. (1983). "Bootstrapping Regression Models with Many Parameters." In *A Festschrift for Erich L. Lehmann*, edited by P. Bickel, K. Doksum, and J. Hodges, pp. 28–48. Belmont, CA: Wadsworth.

Bierens, H. (1982). "Consistent Model Specification Tests." *Journal of Econometrics*, 20, 105–34.

Billingsley, P. (1985). *Probability and Measure (2nd Ed)*. New York: Wiley.

Bivand, R. (1984). "Regression Modeling with Spatial Dependence: An Application of Some Class Selection and Estimation Methods." *Geographical Analysis*, 16, 25–37.

Blommestein, H. (1983). "Specification and Estimation of Spatial Econometric Models: A Discussion of Alternative Strategies for Spatial Economic Modelling." *Regional Science and Urban Economics*, 13, 251–70.

Blommestein, H. (1985). "Elimination of Circular Routes in Spatial Dynamic Regression Equations." *Regional Science and Urban Economics*, 15, 121–130.

Blommestein, H. and P. Nijkamp. (1986). "Testing the Spatial Scale and the Dynamic Structure in Regional Models. A Contribution to Spatial Econometric Specification Analysis." *Journal of Regional Science*, 26, 1–17.

Bodson, P. and D. Peeters. (1975). "Estimation of the Coefficients of a Linear Regression in the Presence of Spatial Autocorrelation: An Application to a Belgian Labour–Demand Function." *Environment and Planning A*, 7, 455–72.

Boots, B. (1982). "Comments on the Use of Eigenfunctions to Measure Structural Properties of Geographic Networks." *Environment and Planning A*, 14, 1063–72.

Boots, B. (1984). "Evaluating Principal Eigenvalues as Measures of Network Structure." *Geographical Analysis*, 16, 270–75.

Boots, B. (1985). "Size Effects in the Spatial Patterning of Nonprincipal Eigenvectors of Planar Networks." *Geographical Analysis*, 17, 74–81.

Boots, B., and K. Tinkler. (1983). "The Interpretation of Non–Principal Eigenfunctions of Urban Structure." *Modeling and Simulation*, 14, 715–19.

Bowden, R. and D. Turkington. (1984). *Instrumental Variables*. London: Cambridge University Press.

Box, G. (1980). "Sampling and Bayes Inference in Scientific Modelling and Robustness." *Journal of the Royal Statistical Society A*, 143, 383–430.

Box, G. and G. Jenkins. (1976). *Time Series Analysis, Forecasting and Control*. San Francisco: Holden Day.

Box, G. and G. Tiao. (1973). *Bayesian Inference in Statistical Analysis*. Reading, MA: Addison–Wesley.

Brandsma, A., and R. Ketellapper. (1979a). "A Biparametric Approach to Spatial Autocorrelation." *Environment and Planning A*, 11, 51–58.

Brandsma, A., and R. Ketellapper. (1979b). "Further Evidence on Alternative Procedures for Testing of Spatial Autocorrelation Among Regression Disturbances." In *Exploratory and Explanatory Analysis in Spatial Data*, edited by C. Bartels and R. Ketellapper, pp. 111–36. Boston: Martinus Nijhoff.

Bretschneider, S., and W. Gorr. (1981). "On the Relationship of Adaptive Filtering Forecasting Models to Simple Brown Smoothing." *Management Science*, 27, 965–9.

Bretschneider, S., and W. Gorr. (1983). "Ad Hoc Model Building Using Time–Varying Parameter Models." *Decision Sciences*, 14, 221–39.

Breusch, T. (1979). "Conflict Among Criteria for Testing Hyptheses: Extensions and Comments." *Econometrica*, 47, 203–7.

Breusch, T. (1980). "Useful Invariance Results for Generalized Regression Models." *Journal of Econometrics*, 13, 327–40.

Breusch, T. (1987). "Maximum Likelihood Estimation of Random Effects Models." *Journal of Econometrics*, 36, 383–9.

Breusch, T., and L. Godfrey. (1981). "A Review of Recent Work on Testing for Auto–Correlation in Dynamic Simultaneous Models." In *Macro–Economic Analysis*, edited by D. Currie, R. Nobay, and D. Peel, pp. 63–100. London: Croom Helm.

Breusch, T., and A. Pagan. (1979). "A Simple Test for Heteroskedasticity and Random Coefficient Variation." *Econometrica*, 47, 1287–94.

Breusch, T., and A. Pagan. (1980). "The Lagrange Multiplier Test and its Applications to Model Specification in Econometrics." *Review of Economic Studies*, 67, 239–53.

Bronars, S., and D. Jansen. (1987). "The Geographic Distribution of Unemployment Rates in the U.S." *Journal of Econometrics*, 36, 251–79.

Brown, R., J. Durbin, and J. Evans. (1975). "Techniques for Testing the Constancy of Regression Relationships over Time." *Journal of the Royal Statistical Society B*, 37, 149–63.

Brown, L, and J–P. Jones. (1985). "Spatial Variation in Migration Processes and Development: A Costa Rican Example of Conventional Modeling Augmented by the Expansion Method." *Demography*, 22, 327–52.

Brueckner, J. (1981). "Testing a Vintage Model of Urban Growth." *Journal of Regional Science*, 21, 23–35.

Brueckner, J. (1985). "A Switching Regression Analysis of Urban Population Densities: Preliminary Results." *Papers, Regional Science Association*, 56, 71−87.

Brueckner, J. (1986). "A Switching Regression Analysis of Urban Population Densities." *Journal of Urban Economics*, 19, 174−89.

Buck, A., and S. Hakim. (1980). "Model Selection in Analyzing Spatial Groups in Regression Analysis." *Geographical Analysis*, 12, 392−98.

Buck, A., and S. Hakim. (1981). "Appropriate Roles for Statistical Decision Theory and Hypothesis Testing in Model Selection." *Regional Science and Urban Economics*, 11, 135−47.

Burridge, P. (1980). "On the Cliff−Ord Test for Spatial Correlation." *Journal of the Royal Statistical Society B*, 42, 107−8.

Burridge, P. (1981). "Testing for a Common Factor in a Spatial Autoregressive Model." *Environment and Planning A*, 13, 795−800.

Burt, S. (1980). "Models of Network Structure." *Annual Review of Sociology*, 6, 79−141.

Buse, A. (1973). "Goodness−of−fit in Generalized Least Squares Estimation." *The American Statistician*, 27, 106−8.

Buse, A. (1979). "Goodness−of−fit in the Seemingly Unrelated Regressions Model, A Generalization." *Journal of Econometrics*, 10, 109−13.

Buse, A. (1982). "The Likelihood Ratio, Wald, and Lagrange Multiplier Tests: An Expository Note." *The American Statistician*, 36, 153−7.

Carbone, R., and W. Gorr. (1978). "An Adaptive Diagnostic Model for Air Quality Management." *Atmospheric Environment*, 12, 1785−91.

Carbone, R., and R. Longini. (1977). "A Feedback Model for Automated Real Estate Assessment." *Management Science*, 24, 241−8.

Casetti, E. (1972). "Generating Models by the Expansion Method: Applications to Geographical Research." *Geographical Analysis*, 4, 81−91.

Casetti, E. (1986). "The Dual Expansion Method: An Application for Evaluating the Effects of Population Growth on Development." *IEEE Transactions on Systems, Man, and Cybernetics*, SMC−16, 29−39.

Casetti, E., and J−P. Jones. (1987). "Spatial Aspects of the Productivity Slowdown: An Analysis of U.S. Manufacturing Data." *Annals, Association of American Geographers*, 77, 76−88.

Casetti, E., and J−P. Jones. (1988). "Spatial Parameter Variation by Orthogonal Trend Surface Expansions: An Application to the Analysis of Welfare Program Participation Rates." *Social Science Research*, 16, 285−300.

Chamberlain, G. (1982). "Multivariate Regression Models for Panel Data." *Journal of Econometrics*, 18, 5−46.

Chamberlain, G. (1984). "Panel Data." In *Handbook of Econometrics, Vol. II*, edited by Z. Griliches and M. Intriligator, pp. 1247−1318. Amsterdam: North Holland.

Charney, A., and C. Taylor. (1983). "Consistent Region−Subregion Econometric Models: A Comparison of Multiarea Methodologies for Arizona and its Major Subcomponent Regions." *International Regional Science Review*, 8, 59−74.

Charney, A., and C. Taylor. (1984). "Decomposition of Ex Ante State Model Forecast Errors." *Journal of Regional Science*, 24, 229−47.

Chesher, A. (1984). "Testing for Neglected Heterogeneity." *Econometrica*, 52, 865−72.

Chesher, A., and I. Jewitt. (1987). "The Bias of a Heteroskedasticity Consistent Covariance Matrix Estimator." *Econometrica*, 55, 1217−22.

Chow, G. (1960). "Tests of Equality Between Sets of Coefficients in Two Linear Regressions." *Econometrica*, 28, 591−605.

Chow, G. (1981). "A Comparison of the Information and Posterior Probability Criteria for Model Selection." *Journal of Econometrics*, 16, 21−33.

Chow, G. (1984). "Random and Changing Coefficient Models." In *Handbook of Econometrics*, Vol. *II*, edited by Z. Griliches and M. Intriligator, pp. 1213−45. Amsterdam: North Holland.

Chow, G., and P. Corsi. (1982). *Evaluating the Reliability of Macro−Economic Models.* New York: Wiley.

Christ, C. (1975). "Judging the Performance of Econometric Models of the U.S. Economy." *International Economic Review*, 16, 54−74.

Chung, K. (1974). *A Course in Probability Theory.* New York: Academic Press.

Clark, I. (1979). *Practical Geostatistics.* London: Applied Science Publishers.

Cliff, A. and J. Ord. (1972). "Testing for Spatial Autocorrelation Among Regression Residuals." *Geographical Analysis*, 4, 267−84.

Cliff, A. and J. Ord. (1973). *Spatial Autocorrelation.* London: Pion.

Cliff, A. and J. Ord. (1981). *Spatial Processes, Models and Applications.* London: Pion.

Cochrane, D., and G. Orcutt. (1949). "Application of Least Squares Regression to Relationships Containing Autocorrelated Error Terms." *Journal of the American Statistical Association*, 44, 32−61.

Consigliere, I. (1981). "The Chow Test with Serially Correlated Errors." *Rivista Internazionale Di Scienze Sociali*, 89, 125−37.

Cook, D., and S. Pocock. (1983). "Multiple Regression in Geographical Mortality Studies, with Allowance for Spatially Correlated Errors." *Biometrics*, 39, 361−71.

Cooley, T., and S. LeRoy. (1981). "Identification and Estimation of Money Demand." *American Economic Review*, 71, 825−44.

Cooley, T., and S. LeRoy. (1985). "Atheoretical Macro−Econometrics: A Critique." *Journal of Monetary Economics*, 16, 283−308.

Cooley, T., and S. LeRoy. (1986). "What Will Take the Con Out Of Econometrics? A Reply to McAleer, Pagan and Volker." *American Economic Review*, 76, 504−7.

Cooley, T., and E. Prescott. (1973). "Systematic (Non−Random) Variation Models Varying Parameter Regression: A Theory and Some Applications." *Annals of Economic and Social Measurement*, 2/4, 463−73.

Cooley, T., and E. Prescott. (1976). "Estimation in the Presence of Stochastic Parameter Variation." *Econometrica*, 44, 167−84.

Corsi, P., R. Pollock, and J. Prakken. (1982). "The Chow Test in the Presence of Serially Correlated Errors." In *Evaluating the Reliability of Macro−Economic Models*, edited by G. Chow and P. Corsi, pp. 163−87. New York: John Wiley.

Costanzo, C. (1982). "QAP II Fortran Program." Discussion Paper, Department of Geography, University of California, Santa Barbara.

Costanzo, C. (1983). "Statistical Inference in Geography: Modern Approaches Spell Better Times Ahead." *The Professional Geographer*, 35, 158−65.

Couclelis, H., and R. Golledge. (1983). "Analytical Research, Positivism and Behavioral Geography." *Annals of the Association of American Geographers*, 73, 331−9.

Cox, D. (1961). "Tests of Separate Families of Hypotheses." *Proceedings of the Fourth Berkeley Symposium on Mathematical Statistics and Probability*, 1, 105−23.

Cox, D. (1962). "Further Results on Tests of Separate Families of Hypotheses." *Journal of the Royal Statistical Society B*, 24, 406−24.

Cox, K., and R. Golledge. (1981). *Behavioral Problems in Geography Revisited.* New York: Methuen.

Cragg, J. (1983). "More Efficient Estimation in the Presence of Heteroskedasticity of Unknown Form." *Econometrica*, 51, 751−63.

Crowder, M. (1976). "Maximum Likelihood Estimation for Dependent Observations." *Journal of the Royal Statistical Society B*, 38, 45−53.

Dacey, M. (1968). "A Review of Measures of Contiguity for Two and K—Color Maps." *In Spatial Analysis: A Reader in Statistical Geography*, edited by B. Berry and D. Marble, pp. 479—95. Englewood Cliffs, N.J.: Prentice—Hall.

Dastoor, N. (1981). "A Note on the Interpretation of the Cox Procedure for Non—Nested Hypotheses." *Economics Letters*, 8, 113—20.

Davidson, R. and J. MacKinnon. (1981). "Several Tests for Model Specification in the Presence of Alternative Hypotheses." *Econometrica*, 49, 781—93.

Davidson, R. and J. MacKinnon. (1982). "Some Non—Nested Hypothesis Tests and the Relations Among Them." *Review of Economic Studies*, 49, 551—65.

Davidson, R. and J. MacKinnon. (1983). "Small Sample Properties of Alternative Forms of the Lagrange Multiplier Test." *Economics Letters*, 12, 269—75.

Davidson, R. and J. MacKinnon. (1984). "Model Specification Tests Based on Artificial Linear Regressions." *International Economic Review*, 25, 485—502.

Davidson, R. and J. MacKinnon. (1985a). "Heteroskedasticity—Robust Tests in Regression Directions." *Annales De L'INSEE*, 59/60, 183—217.

Davidson, R., and J. MacKinnon. (1985b). "The Interpretation of Test Statistics." *Canadian Journal of Economics*, 18, 38—57.

Dhrymes, P. (1981). *Distributed Lags: Problems of Estimation and Formulation.* Amsterdam: North—Holland.

Dhrymes, P., E. Howrey, S. Hymans, J. Kmenta, E. Leamer, R. Quandt, J. Ramsey, H. Shapiro, and V. Zarnowitz. (1972). "Criteria for Evaluation of Econometric Models." *Annals of Economic and Social Measurement*, 1, 291—324.

Dielman, T. (1983). "Pooled Cross—Sectional and Time Series Data: A Survey of Current Statistical Methodology." *The American Statistician*, 37, 111—22.

Diggle, P. (1983). *Statistical Analysis of Spatial Point Patterns.* New York: Academic Press.

Domowitz, I., and H. White. (1982). "Misspecified Models with Dependent Observations." *Journal of Econometrics*, 20, 35—58.

Doob, J. (1953). *Stochastic Processes.* New York: Wiley.

Doran, H., and W. Griffiths. (1983). "On the Relative Efficiency of Estimators which Include the Initial Observations in the Estimation of Seemingly Unrelated Regressions with First—Order Autoregressive Disturbances." *Journal of Econometrics*, 23, 165—91.

Doreian, P. (1974). "On the Connectivity of Social Networks." *Journal of Mathematical Sociology*, 3, 245—58.

Doreian, P. (1980). "Linear Models with Spatially Distributed Data: Spatial Disturbances or Spatial Effects?" *Sociological Methods and Research*, 9, 29—60.

Doreian, P. (1981). "Estimating Linear Models with Spatially Distributed Data." In *Sociological Methodology 1980*, edited by S. Leinhardt, pp. 359—88. San Francisco: Jossey—Bass.

Doreian, P. (1982). "Maximum Likelihood Methods for Linear Models: Spatial Effect and Spatial Disturbance Terms." *Sociological Methods and Research*, 13, 243—69.

Doreian, P., K. Teuter, and C—H Wang. (1984). "Network Autocorrelation Models." *Sociological Methods and Research*, 13, 155—200.

Dreze, J. (1972). "Econometrics and Decision Theory." *Econometrica*, 40, 1—17.

Dufour, J—M. (1982). "Generalized Chow Tests for Structural Change: A Coordinate—Free Approach." *International Economic Review*, 23, 565—75.

Duncan, G. (1983). "Estimation and Inference for Heteroscedastic Systems of Equations." *International Economic Review*, 24, 559—66.

Durbin, J. (1960). "Estimation of Parameters in Time—Series Regression Models." *Journal of the Royal Statistical Society B*, 22, 139—53.

Edgington, E. (1969). *Statistical Inference: The Distribution—Free Approach.* New York: McGraw—Hill.

Edgington, E. (1980). *Randomization Tests*. New York: Marcel Dekker.

Edlefsen, L. and S. Jones. (1986). *GAUSS Programming Language Manual*. Seattle, WA: Aptech Systems Inc.

Efron, B. (1979a). "Computers and the Theory of Statistics: Thinking the Unthinkable." *SIAM Review*, 21, 460−80.

Efron, B. (1979b). "Bootstrap Methods: Another Look at the Jackknife." *The Annals of Statistics*, 7, 1−26.

Efron, B. (1982). *The Jackknife, the Bootstrap and Other Resampling Plans*. Philadelphia: Society for Industrial and Applied Mathematics.

Efron, B. and G. Gong. (1983). "A Leisurely Look at the Bootstrap, the Jackknife, and Cross−Validation." *The American Statistician*, 37, 36−48.

Engle, R. (1982). "A General Approach to Lagrange Multiplier Model Diagnostics." *Journal of Econometrics*, 20, 83−104.

Engle, R. (1984). "Wald, Likelihood Ratio, and Lagrange Multiplier Tests in Econometrics." In *Handbook of Econometrics, Vol. II*, edited by Z. Griliches and M. Intriligator, pp. 775−826. Amsterdam: North Holland.

Epps, T., and M. Epps. (1977). "The Robustness of Some Standard Tests for Autocorrelation and Heteroskedasticity when Both Problems are Present." *Econometrica*, 45, 745−53.

Ericsson, N. (1983). "Asymptotic Properties of Instrumental Variables Statistics for Testing Non−Nested Hypotheses." *Review of Economic Studies*, 50, 287−304.

Evans, G., and N. Savin. (1982a). "Conflict Among the Criteria Revisited; the W, LR and LM Tests." *Econometrica*, 50, 737−48.

Evans, G., and N. Savin. (1982b). "Conflict Among Testing Procedures in a Linear Regression Model with Lagged Dependent Variables." In *Advances in Econometrics*, edited by W. Hildenbrand, pp. 263−83. London: Cambridge University Press.

Fienberg, S., and A. Zellner. (1975). *Studies in Bayesian Econometrics and Statistics in Honor of Leonard J. Savage*. Amsterdam: North Holland.

Fik, T. (1988). "Competing Central Places and the Spatially Autocorrelated Seemingly Unrelated Regression System." *The Annals of Regional Science*, 22 (forthcoming).

Fischer, M. (1984). "Theory and Testing in Empirical Sciences." In *Recent Developments in Spatial Data Analysis: Methodology, Measurement, Models*, edited by G. Bahrenberg, M. Fischer, and P. Nijkamp, pp. 51−71. Aldershot: Gower.

Fisher, F. (1966). *The Identification Problem in Econometrics*. New York: McGraw Hill.

Fisher, F. (1970). "Tests of Equality Between Sets of Coefficients in Two Linear Regressions: An Expository Note." *Econometrica*, 38, 361−66.

Fisher, G. (1983). "Tests for Two Separate Regressions." *Journal of Econometrics*, 21, 117−32.

Fisher, G. and M. McAleer. (1979). "On the Interpretation of the Cox Test in Econometrics." *Economics Letters*, 4, 145−50.

Fisher, G. and M. McAleer. (1981). "Alternative Procedures and Associated Tests of Significance for Non−Nested Hypotheses." *Journal of Econometrics*, 16, 103−19.

Fisher, W. (1971). "Econometric Estimation with Spatial Dependence." *Regional and Urban Economics*, 1, 19−40.

Folmer, H. (1986). *Regional Economic Policy: Measurement of its Effect*. Dordrecht: Martinus Nijhoff.

Folmer, H. and M. Fischer. (1984). "Bootstrapping in Spatial Analysis." Paper presented at the Symposium of the IGU Working Group on Systems Analysis and Mathematical Models, Besancon, France.

Folmer, H. and P. Nijkamp. (1984). "Linear Structural Equation Models with Latent Variables and Spatial Correlation." In *Recent Developments in Spatial Data Analysis*, edited by G. Bahrenberg, M. Fischer and P. Nijkamp, pp. 163−70. Aldershot: Gower.

Folmer, H. and G. van der Knaap. (1981). "A Linear Structural Equation Approach to Cross—Sectional Models with Lagged Variables." *Environment and Planning A*, 13, 1529—37.

Foster, S., and W. Gorr. (1983). "Adaptive Filtering Approaches to Spatial Modeling." *Modeling and Simulation*, 14, 745—50.

Foster, S., and W. Gorr. (1984). "Spatial Adaptive Filtering." *Modeling and Simulation*, 15, 29—34.

Foster, S., and W. Gorr. (1986). "An Adaptive Filter for Estimating Spatially—Varying Parameters: Application to Modeling Police Hours Spent in Response to Calls for Service." *Management Science*, 32, 878—89.

Fotheringham, S., and D. Knudsen. (1986). *Goodness—of—fit Statistics in Geographic Research*. Norwich: Geo Abstracts.

Freedman, D. (1981). "Bootstrapping Regression Models." *The Annals of Statistics*, 9, 1218—1228.

Freedman, D. and S. Peters. (1984a). "Bootstrapping a Regression Equation: Some Empirical Results." *Journal of the American Statistical Association*, 79, 97—106.

Freedman, D. and S. Peters. (1984b). "Bootstrapping an Econometric Model: Some Empirical Results." *Journal of Business and Economic Statistics*, 2, 150—158.

Frisch, R. (1981). "From Utopian Theory to Practical Applications: The Case of Econometrics." *American Economic Review*, 71, 1—7.

Gaile, G. and C. Wilmott. (1984). *Spatial Statistics and Models*. Boston: Reidel.

Gale, N., L. Hubert, W. Tobler, and R. Golledge. (1983). "Combinatorial Procedures for the Analysis of Alternative Models: An Example from Interregional Migration." *Papers, Regional Science Association*, 53, 105—15.

Gallant, A., and D. Jorgenson. (1979). "Statistical Inference for a System of Non—Linear Implicit Equations in the Context of Instrumental Variables Estimation." *Journal of Econometrics*, 11, 275—302.

Garrison, W., and D. Marble. (1964). "Factor—Analytic Study of the Connectivity of a Transportation Network." *Papers, Regional Science Association*, 12, 231—38.

Gatrell, A. (1977). "Complexity and Redundancy in Binary Maps." *Geographical Analysis*, 9, 29—41.

Gatrell, A. (1979). "Autocorrelation in Spaces." *Environment and Planning A*, 11, 507—16.

Gaver, K., and M. Geisel. (1974). "Discriminating Among Alternative Models: Bayesian and Non—Bayesian Methods." In *Frontiers in Econometrics*, edited by P. Zarembka, pp. 49—77. New York: Academic Press.

Geary, R. (1954). "The Contiguity Ratio and Statistical Mapping." *The Incorporated Statistician*, 5, 115—45.

Geisel, M. (1975). "Bayesian Comparisons of Simple Macroeconomic Models. In *Studies in Bayesian Econometrics and Statistics in Honor of Leonard J. Savage*, edited by S. Fienberg and A. Zellner, pp. 227—56. Amsterdam: North Holland.

Geisser, S. (1975). "The Predictive Sample Reuse Method with Applications." *Journal of the American Statistical Association*, 70, 320—28.

Geisser, S., and W. Eddy. (1979). "A Predictive Approach to Model Selection." *Journal of the American Statistical Association*, 74, 153—60.

Getis, A. and B. Boots. (1978). *Models of Spatial Processes*. London: Cambridge University Press.

Ghali, M. (1977). "Pooling as a Specification Error: A Comment." *Econometrica*, 45, 755—7.

Giles, D., and M. Beattie. (1987). "Autocorrelation Pre—Test Estimation in Models with a Lagged Dependent Variable." In *Specification Analysis in the Linear Model*, edited by M. King and D. Giles, pp. 99—116. London: Routledge and Kegan Paul.

Glejser, H. (1969). "A New Test for Heteroskedasticity." *Journal of the American Statistical Association*, 64, 316−23.

Glick, B. (1982). "A Spatial Rank−Order Correlation Measure." *Geographical Analysis*, 14, 177−81.

Godfrey, L. (1978). "Testing for Multiplicative Heteroskedasticity." *Journal of Econometrics*, 8, 227−36.

Godfrey, L. (1983). "Testing Non−Nested Models After Estimation by Instrumental Variables or Least Squares." *Econometrica*, 51, 355−65.

Godfrey, L. (1984). "On the Uses of Misspecification Checks and Tests of Non−Nested Hypotheses in Empirical Econometrics." *Economic Journal*, 94, 69−81.

Godfrey, L. (1987). "Discriminating Between Autocorrelation and Misspecification in Regression Analysis: An Alternative Test Strategy." *The Review of Economics and Statistics*, 69, 128−34.

Godfrey, L. and M. Pesaran. (1983). "Tests of Non−Nested Regression Models: Small Sample Adjustments and Monte Carlo Evidence." *Journal of Econometrics*, 21, 133−54.

Goldfeld, S. and R. Quandt. (1972). *Nonlinear Methods in Econometrics*. Amsterdam: North Holland.

Goldfeld, S. and R. Quandt. (1973). "The Estimation of Structural Shifts by Switching Regressions." *Annals of Economic and Social Measurement*, 2/4, 475−85.

Goldfeld, S. and R. Quandt. (1976). "Techniques for Estimating Switching Regressions." In *Studies in Nonlinear Estimation*, edited by S. Goldfeld and R. Quandt, pp. 3−35. Cambridge, MA: Ballinger.

Golledge, R. (1980). "A Behavioral View of Mobility and Migration Research." *The Professional Geographer*, 32, 14−21.

Golledge, R., L. Hubert, and G. Richardson. (1981). "The Comparison of Related Data Sets: Examples from Multidimensional Scaling and Cluster Analysis." *Papers, Regional Science Association*, 48, 57−66.

Gould, P. (1967). "On the Geographical Interpretation of Eigenvalues." *Transactions, Institute of British Geographers*, 42, 53−92.

Gould, P. (1970). "Is Statistix Inferens the Geographical Name for a Wild Goose?" *Economic Geography*, 46, 439−48.

Gould, P. (1981). "Letting the Data Speak for Themselves." *Annals of the Association of American Geographers*, 71, 166−76.

Gourieroux, C., A. Monfort, A. Trognon. (1983). "Testing Nested or Non−Nested Hypotheses." *Journal of Econometrics*, 21, 83−115.

Gourieroux, C., A. Monfort, A. Trognon. (1984a). "Pseudo Maximum Likelihood Methods: Theory." *Econometrica*, 52, 681−700.

Gourieroux, C., A. Monfort, A. Trognon. (1984b). "Pseudo Maximum Likelihood Methods: Applications to Poisson Models." *Econometrica*, 52, 701−20.

Granger, C. (1969). "Spatial Data and Time Series Analysis." In *Studies in Regional Science, London Papers in Regional Science*, edited by A. Scott, pp. 1−24. London: Pion.

Granger, C., and P. Newbold. (1974). "Spurious Regressions in Econometrics." *Journal of Econometrics*, 2, 111−20.

Greenberg, E., and C. Webster. (1983). *Advanced Econometrics, A Bridge to the Literature*. New York: Wiley.

Greene, D., and J. Barnbock. (1978). "A Note on Problems in Estimating Urban Density Functions." *Journal of Urban Economics*, 5, 285−90.

Greenwood, M., G. Hunt, J. McDowell. (1986). "Migration and Empoyment Change: Empirical Evidence on the Spatial and Temporal Dimensions of the Linkage." *Journal of Regional Science*, 26, 223−234.

Grether, D., and G. Maddala. (1973). "Errors in Variables and Serially Correlated Disturbances in Distributed Lag Models." *Econometrica*, 41, 255–62.

Griffith, D. (1978). "A Spatially Adjusted ANOVA Model." *Geographical Analysis*, 10, 296–301.

Griffith, D. (1980). "Towards a Theory of Spatial Statistics." *Geographical Analysis*, 12, 325–39.

Griffith, D. (1983). "The Boundary Value Problem in Spatial Statistical Analysis." *Journal of Regional Science*, 23, 377–87.

Griffith, D. (1984). "Measurement of the Arrangement Property of a System of Areal Units Generated by Partitioning a Planar Surface." In *Recent Developments in Spatial Data Analysis*, edited by G. Bahrenberg, M. Fischer, and P. Nijkamp, pp. 191–99. Aldershot: Gower.

Griffith, D. (1985). "An Evaluation of Correction Techniques for Boundary Effects in Spatial Statistical Analysis: Contemporary Methods." *Geographical Analysis*, 17, 81–8.

Griffith, D. (1987). "Towards a Theory of Spatial Statistics: Another Step Forward." *Geographical Analysis*, 19, 69–82.

Griffith, D. (1988a). "A Reply to: Some Comments on Correction Techniques for Boundary Effects and Missing Value Techniques." *Geographical Analysis*, 20, 70–5.

Griffith, D. (1988b). "Estimating Spatial Autoregressive Model Parameters with Commercial Statistical Packages." *Geographical Analysis*, 20 (in press).

Griffith, D., and C. Amrhein. (1983). "An Evaluation of Correction Techniques for Boundary Effects in Spatial Statistical Analysis: Traditional Methods." *Geographical Analysis*, 15, 352–60.

Griffiths, W., and P. Beesley. (1984). "The Small Sample Properties of Some Preliminary Test Estimators in a Linear Model with Autocorrelated Errors." *Journal of Econometrics*, 25, 49–61.

Guilkey, D., and P. Schmidt. (1973). "Estimation of Seemingly Unrelated Regressions with Vector Autoregressive Errors." *Journal of the American Statistical Association*, 68, 642–7.

Haggett, P. (1980). "Boundary Problems in Statistical Geography." In *Die Bedeutung von Grenzen in der Geographie*, edited by H. Kishimoto, pp. 59–67. Zurich: Kummerley and Frey.

Haggett, P. (1981). "The Edges of Space." In *European Progress in Spatial Analysis*, edited by R. Bennett, pp. 51–70. London: Pion.

Haining, R. (1977). "Model Specification in Stationary Random Fields." *Geographical Analysis*, 9, 107–29.

Haining, R. (1978a). "Estimating Spatial Interaction Models." *Environment and Planning A*, 10, 305–320.

Haining, R. (1978b). "Interaction Modelling on Central Place Lattices." *Journal of Regional Science*, 18, 217–228.

Haining, R. (1978c). "The Moving Average Model for Spatial Interaction." *Transactions, Institute of British Geographers*, 3, 202–25.

Haining, R. (1978d). *Specification and Estimation Problems in Models of Spatial Dependence*. Evanston, Ill.: Northwestern University Studies in Geography no. 24.

Haining, R. (1979). "Statistical Tests and Process Generators for Random Field Models." *Geographical Analysis*, 11, 45–64.

Haining, R. (1984). "Testing a Spatial Interacting Markets Hypothesis." *The Review of Economics and Statistics*, 66, 576–83.

Haining, R. (1986a). "Income Diffusion and Regional Economics." In *Transformations Through Space and Time*, edited by D. Griffith and R. Haining, pp. 59–80. Dordrecht: Martinus Nijhoff.

Haining, R. (1986b). "Spatial Models and Regional Science: A Comment on Anselin's Paper and Research Directions." *Journal of Regional Science*, 26, 793—8.

Haitovsky, Y., and G. Treyz. (1970). "The Analysis of Econometric Forecasting Error." *American Statistical Association, Proceedings of the Business and Economics Section*, 17, 502—6.

Haitovsky, Y., and G. Treyz. (1972). "Forecasts with Quarterly Macroeconomic Models: Equation Adjustments and Benchmark Predictions, the U.S. Experience." *The Review of Economics and Statistics*, 54, 317—25.

Hamilton, J., and R. Jensen. (1983). "Summary Measures of Interconnectedness for Input—Output Models." *Environment and Planning A*, 15, 55—65.

Hanham, R., M. Hohn, and J. Bohland. (1984). "Kriging Spatial Data: Application to the Distribution of Elderly in the U.S." *Modeling and Simulation*, 15, 35—39.

Hannan, E., and B. Quinn. (1979). "The Determination of the Order of an Autoregression." *Journal of the Royal Statistical Society B*, 41, 190—95.

Hansen, L. (1982). "Large Sample Properties of Generalized Method of Moments Estimators." *Econometrica*, 50, 1029—54.

Harrison, M., and B. Mc Cabe. (1975). "Autocorrelation with Heteroskedasticity: A Note on the Robustness of the Durbin—Watson, Geary and Henshaw Tests." *Biometrika*, 62, 214—6.

Harvey, A. (1976). "Estimating Regression Models with Multiplicative Heteroskedasticity." *Econometrica*, 44, 461—6.

Hausman, J. (1978). "Specification Tests in Econometrics." *Econometrica*, 46, 1251—70.

Hausman, J. (1984). "Specification and Estimation of Simultaneous Equation Models." In *Handbook of Econometrics*, edited by Z. Griliches and M. Intrilligator, pp. 392—448. Amsterdam: North Holland.

Heijmans, R., and J. Magnus. (1986a). "On the First—Order Efficiency and Asymptotic Normality of Maximum Likelihood Estimators Obtained from Dependent Observations." *Statistica Neerlandica*, 40, 169—88.

Heijmans, R., and J. Magnus. (1986b). "Consistent Maximum—Likelihood Estimation with Dependent Observations: The General (Non—Normal) Case and the Normal Case." *Journal of Econometrics*, 32, 253—85.

Heijmans, R., and J. Magnus. (1986c). "Asymptotic Normality of Maximum Likelihood Estimators Obtained from Normally Distributed but Dependent Observations." *Econometric Theory*, 2, 374—412.

Hendry, D. (1980). "Econometrics — Alchemy or Science? *Economica*, 47, 387—406.

Hendry, D., and G. Mizon. (1978). "Serial Correlation as a Convenient Simplification, not a Nuisance: A Comment on a Study of the Demand for Money by the Bank of England." *Economic Journal*, 88, 549—63.

Hendry, D., and F. Srba. (1977). "The Properties of Autoregressive Instrumental Variables Estimators in Dynamic Systems." *Econometrica*, 45, 969—90.

Hepple, L. (1976). "A Maximum Likelihood Model for Econometric Estimation with Spatial Series." In *Theory and Practice in Regional Science, London Papers in Regional Science 6*, edited by I. Masser, pp. 90—104. London: Pion.

Hepple, L. (1979). "Bayesian Analysis of the Linear Model with Spatial Dependence." In *Exploratory and Explanatory Statistical Analysis of Spatial Data*, edited by C. Bartels and R. Ketellapper, pp. 179—99. Boston: Martinus Nijhoff.

Hildreth, C., and J. Houck. (1968). "Some Estimators for a Linear Model with Random Coefficients." *Journal of the American Statistical Association*, 63, 584—95.

Holly, A. (1982). "A Remark on Hausman's Specification Test." *Econometrica*, 50, 749—59.

Honda, Y. (1982). "On Tests of Equality Between Sets of Coefficients in Two Linear Regressions when Disturbance Variances are Unequal." *The Manchester School of Economic and Social Studies*, 50, 116–25.

Hooper, P. and G. Hewings. (1981). "Some Properties of Space–Time Processes." *Geographical Analysis*, 13, 203–23.

Hope, A. (1968). "A Simplified Monte Carlo Significance Test Procedure." *Journal of the Royal Statistical Society*, 30B, 582–98.

Hordijk, L. (1974). "Spatial Correlation in the Disturbance of a Linear Interregional Model." *Regional Science and Urban Economics*, 4, 117–40.

Hordijk, L. (1979). "Problems in Estimating Econometric Relations in Space." *Papers, Regional Science Association*, 42, 99–115.

Hordijk, L. and P. Nijkamp. (1977). "Dynamic Models of Spatial Autocorrelation." *Environment and Planning A*, 9, 505–519.

Hordijk, L. and P. Nijkamp. (1978). "Estimation of Spatio–Temporal Models: New Directions Via Distributed Lags and Markov Schemes." In *Spatial Interaction Theory and Planning Models*, edited by A. Karlquist, L. Lundquist, F. Snickars, and J. Weibull, pp. 177–199. Amsterdam: North Holland.

Hordijk, L. and J. Paelinck. (1976). "Some Principles and Results in Spatial Econometrics." *Recherches Economiques de Louvain*, 42, 175–97.

Horowitz, J. (1982). "Specification Tests for Probabilistic Choice Models." *Transportation Research A*, 16, 383–94.

Horowitz, J. (1983). "Statistical Comparison of Non–Nested Probabilistic Discrete Choice Models." *Transportation Science*, 17, 319–50.

Horowitz, J. (1985). "Testing Probabilistic Discrete Choice Models of Travel Demand by Comparing Predicted and Observed Aggregate Choice Shares." *Transportation Research B*, 19, 17–38.

Horowitz, J. (1987). "Specification Tests for Probabilistic Choice Models of Consumer Behavior." In *Behavioral Modeling in Geography and Planning*, edited by R. Golledge and H. Timmermans, pp. 124–37. London: Croom Helm.

Hsiao, C. (1985). "Benefits and Limitations of Panel Data." *Econometric Reviews*, 4, 121–74.

Hsiao, C. (1986). *Analysis of Panel Data*. Cambridge: Cambridge University Press.

Hsieh, D. (1983). "A Heteroskedasticity–Consistent Covariance Matrix Estimator for Time Series Regressions." *Journal of Econometrics*, 22, 281–90.

Huber, P. (1972). "Robust Statistics: A Review." *Annals of Mathematical Statistics*, 43, 1041–67.

Huber, P. (1981). *Robust Statistics*. New York: Wiley.

Hubert, L. (1984). "Statistical Applications of Linear Assignment." *Psychometrika*, 49, 449–73.

Hubert, L. (1985). "Combinatorial Data Analysis: Association and Partial Association." *Psychometrika*, 50, 449–67.

Hubert, L., and R. Golledge. (1981). "A Heuristic Method for the Comparison of Related Structures." *Journal of Mathematical Psychology*, 23, 214–26.

Hubert, L., and R. Golledge. (1982a). "Measuring Association Between Spatially Defined Variables: Tjostheim's Index and Some Extensions." *Geographical Analysis*, 14, 273–78.

Hubert, L., and R. Golledge. (1982b). "Comparing Rectangular Data Matrices." *Environment and Planning A*, 14, 1087–95.

Hubert, L., R. Golledge, and C. Costanzo. (1981). "Generalized Procedures for Evaluating Spatial Autocorrelation." *Geographical Analysis*, 13, 224–33.

Hubert, L., R. Golledge, C. Costanzo, and N. Gale. (1985). "Measuring Association Between Spatially Defined Variables: An Alternative Procedure." *Geographical Analysis*, 27, 36–46.

Hwang, H–S. (1981). "Test of Independence Between a Subset of Stochastic Regressors and Disturbances." *International Economic Review*, 21, 749–60.

Ibragimov, I. and Y. Linnik. (1971). *Independent and Stationary Sequences of Random Variables*. The Hague: Wolters–Noordhoff.

Isard, W. (1956). *Location and Space Economy*. Cambridge: MIT Press.

Isard, W. (1969). *General Theory*. Cambridge: MIT Press.

Isard, W., and P. Liossatos. (1979). *Spatial Dynamics and Optimal Space–Time Development*. Amsterdam: North Holland.

Jarque, C., and A. Bera. (1980). "Efficient Tests for Normality, Homoscedasticity and Serial Independence of Regression Residuals." *Economics Letters*, 6, 255–9.

Jayatissa, W. (1977). "Tests of Equality Between Sets of Coefficients in Two Linear Regressions When Disturbance Variances are Unequal." *Econometrica*, 45, 1291–2.

Jeffreys, H. (1967). *Theory of Probability (3rd Ed.)*. London: Oxford University Press.

Johnson, L. (1975). "A Note on Testing for Intraregional Economic Homogeneity." *Journal of Regional Science*, 15, 365–69.

Johnson, S. and J. Kau. (1980). "Urban Spatial Structure: An Analysis with a Varying Coefficient Model." *Journal of Urban Economics*, 7, 141–54.

Johnston, J. (1984). *Econometric Methods*. New York: McGraw–Hill.

Johnston, R. (1984). "Quantitative Ecological Analysis in Human Geography: An Evaluation of Four Problem Areas." In *Recent Developments in Spatial Data Analysis: Methodology, Measurement, Models*, edited by G. Bahrenberg, M. Fischer, and P. Nijkamp, pp. 131–41. Aldershot: Gower.

Jones, J–P. (1983). "Parameter Variation Via the Expansion Method with Tests for Autocorrelation." *Modeling and Simulation*, 14, 853–57.

Jones, J–P. (1984). *Spatial Parameter Variation in Models of AFDC Participation: Analyses Using the Expansion Method*. Unpublished Doctoral Dissertation, Department of Geography, The Ohio State University.

Judge, G., and M. Bock. (1978). *The Statistical Implications of Pre–Test and Stein–Rule Estimators in Econometrics*. Amsterdam: North Holland.

Judge, G., and T. Yancey. (1986). *Improved Methods of Inference in Econometrics*. Amsterdam: North Holland.

Judge, G., W. Griffiths, R. Hill, H. Lutkepohl. and T–C Lee. (1985). *The Theory and Practice of Econometrics, 2nd ed*. New York: Wiley.

Kadane, J., and J. Dickey. (1980). "Bayesian Decision Theory and the Simplification of Models." In *Evaluation of Econometric Models*, edited by J. Kmenta and J. Ramsey, pp. 245–68. New York: Academic Press.

Kadane, J., J. Dickey, R. Winkler, W. Smith, and S. Peters. (1980). "Interactive Elicitation of Opinion for a Normal Linear Model." *Journal of the American Statistical Association*, 75, 845–54.

Karlin, S., and H. Taylor. (1975). *A First Course in Stochastic Processes*. New York: Academic Press.

Kariya, T. (1981). "Tests for the Independence Between Two Seemingly Unrelated Regression Equations." *The Annals of Statistics*, 9, 381–90.

Kau, J., and C. Lee. (1977). "A Random Coefficient Model to Estimate a Stochastic Density Gradient." *Regional Science and Urban Economics*, 7, 169–77.

Kau, J., C. Lee, and R. Chen. (1983). "Structural Shifts in Urban Population Density Gradients: An Empirical Investigation." *Journal of Urban Economics*, 13, 364–77.

Kau, J., C. Lee, and C. Sirmans. (1986). *Urban Econometrics: Model Developments and Empirical Results*. Greenwich, Conn.: JAI Press.

Kennedy, P. (1985). *A Guide to Econometrics, 2nd ed*. Cambridge, MA: MIT Press.

Keynes, J. (1939). "Professor Tinbergen's Method." *Economic Journal*, 49, 558–68.

Kiefer, N. (1978). "Discrete Parameter Variation: Efficient Estimation of a Switching Regression Model." *Econometrica*, 46, 427–34.

Kinal, T., and K. Lahiri. (1984). "A Note on Selection of Regressors." *International Economic Review*, 25, 625–29.

King, M. (1981). "A Small Sample Property of the Cliff–Ord Test for Spatial Correlation." *Journal of the Royal Statistical Society B*, 43, 263–4.

King, M. (1987). "Testing for Autocorrelation in Linear Regression Models: A Survey." In *Specification Analysis in the Linear Model*, edited by M. King and D. Giles, pp. 19–73. London: Routledge and Kegan Paul.

King, M. and M. Evans. (1985). "The Durbin–Watson Test and Cross–Sectional Data." *Economics Letters*, 18, 31–34.

King, M., and M. Evans. (1986). "Testing for Block Effects in Regression Models Based on Survey Data." *Journal of the American Statistical Association*, 81, 677–79.

King, M., and D. Giles. (1984). "Autocorrelation Pre–Testing in the Linear Model: Estimation, Testing and Prediction." *Journal of Econometrics*, 25, 35–48.

Kiviet, J. (1985). "Model Selection Test Procedures in a Single Linear Equation of a Dynamic Simultaneous System and their Defects in Small Samples." *Journal of Econometrics*, 28, 327–62.

Kiviet, J. (1986). "On the Rigour of Some Misspecification Tests for Modelling Dynamic Relationships." *Review of Economic Studies*, 53, 241–61.

Klein, R., and S. Brown. (1984). "Model Selection when there is 'Minimal' Prior Information." *Econometrica*, 52, 1291–1312.

Kmenta, J. (1971). *Elements of Econometrics*. New York: Macmillan.

Kmenta, J., and J. Ramsey. (1980). *Evaluation of Econometric Models*. New York: Academic Press.

Knudsen, D. (1987). "Computer–Intensive Significance–Testing Procedures." *The Professional Geographer*, 39, 208–14.

Knudsen, D., and S. Fotheringham. (1986). "Matrix Comparison, Goodness–of–fit, and Spatial Interaction Modeling." *International Regional Science Review*, 10, 127–47.

Koenker, R. (1982). "Robust Methods in Econometrics." *Econometric Reviews*, 1, 213–55.

Koenker, R. and G. Bassett. (1982). "Robust Tests for Heteroskedasticity Based on Regression Quantiles." *Econometrica*, 50, 43–61.

Kooijman, S. (1976). "Some Remarks on the Statistical Analysis of Grids, Especially with Respect to Ecology." *Annals of Systems Research*, 5, 113–132.

Krasker, W. (1980). "The Role of Bounded–Influence Estimation in Model Selection." *Journal of Econometrics*, 16, 131–8.

Krasker, W., and R. Welsch. (1982). "Efficient Bounded–Influence Regression Estimation." *Journal of the American Statistical Association*, 77, 595–604.

Krasker, W., and R. Welsch. (1985). "Resistant Estimation for Simultaneous–Equations Models Using Weighted Instrumental Variables." *Econometrica*, 53, 1475–88.

Lahiri, K., and D. Egy. (1981). "Joint Estimation and Testing for Functional Form and Heteroskedasticity." *Journal of Econometrics*, 15, 299–307.

Leamer, E. (1974). "False Models and Post–Data Model Construction." *Journal of the American Statistical Association*, 69, 122–31.

Leamer, E. (1978). *Specification Searches: Ad Hoc Inference with Nonexperimental Data*. New York: Wiley.

Leamer, E. (1979). "Information Criteria for Choice of Regression Models: A Comment." *Econometrica*, 47, 507–10.

Leamer, E. (1983). "Let's Take the Con Out Of Econometrics." *American Economic Review*, 73, 31–43.

Leamer, E., and H. Leonard. (1983). "Reporting the Fragility of Regression Estimates." *Review of Economics and Statistics*, 65, 306–17.

Lee, D. (1973). "Requiem for Large–Scale Models." *Journal of the American Institute of Planners*, 39, 163–78.

Lele, S., and K. Ord. (1986). "Besag's Pseudo–Likelihood: Some Optimality Results." Technical Report and Preprints no. 66, Department of Statistics, The Pennsylvania State University.

Lillard, L., and Y. Weiss. (1979). "Components of Variation in Panel Earnings Data: American Scientists 1960–70." *Econometrica*, 47, 437–54.

Lillard, L., and R. Willis. (1978). "Dynamic Aspects of Earning Mobility." *Econometrica*, 46, 985–1012.

Lin, A–L. (1985). "A Note on Testing for Regional Homogeneity of a Parameter." *Journal of Regional Science*, 25, 129–35.

Liviatan, N. (1963). "Consistent Estimation of Distributed Lags." *International Economic Review*, 4, 44–52.

Loftin, C. and S. Ward. (1983). "A Spatial Autocorrelation Model of the Effects of Population Density on Fertility." *American Sociological Review*, 48, 121–28.

Lovell, M. (1983). "Data Mining." *The Review of Economics and Statistics*, 65, 1–12.

MacKinnon, J. (1983). "Model Specification Tests Against Non–Nested Alternatives." *Econometric Reviews*, 2, 85–110.

MacKinnon, J. and H. White. (1985). "Some Heteroskedasticity–Consistent Covariance Matrix Estimators with Improved Finite Sample Properties." *Journal of Econometrics*, 29, 305–25.

MacKinnon, J., H. White, and R. Davidson. (1983). "Tests for Model Specification in the Presence of Alternative Hypotheses: Some Further Results." *Journal of Econometrics*, 21, 53–70.

MaCurdy, T. (1982). "The Use of Time Series Processes to Model the Error Structure of Earnings in a Longitudinal Data Analysis." *Journal of Econometrics*, 18, 83–114.

Maddala, G. (1971). "The Use of Variance Components Models in Pooling Cross Section and Time Series Data." *Econometrica*, 39, 341–58.

Maddala, G. (1977). *Econometrics*. New York: McGraw–Hill.

Maddala, G. (1983). *Limited–Dependent and Qualitative Variables in Econometrics*. New York: Academic Press.

Maeshiro, A. (1980). "New Evidence on the Small Properties of Estimators of SUR Models with Autocorrelated Disturbances – Things Done Halfway May Not Be Done Right." *Journal of Econometrics*, 12, 177–87.

Magdalinos, M. (1985). "Selecting the Best Instrumental Variables Estimator." *Review of Economic Studies*, 52, 473–85.

Magnus, J. (1978). "Maximum Likelihood Estimation of the GLS Model with Unknown Parameters in the Disturbance Covariance Matrix." *Journal of Econometrics*, 7, 281–312. (Corrigenda, *Journal of Econometrics*, 10, 261)

Magnus, J. (1982). "Multivariate Error Components Analysis of Linear and Nonlinear Regression Models by Maximum Likelihood." *Journal of Econometrics*, 19, 239–85.

Magnus, J., and H. Neudecker. (1985). "Matrix Differential Calculus with Applications to Simple, Hadamard, and Kronecker Products." *Journal of Mathematical Psychology*, 29, 474–492.

Magnus, J., and H. Neudecker. (1986). "Symmetry, 0–1 Matrices and Jacobians: A Review." *Econometric Theory*, 2, 157–90.

Malinvaud, E. (1981). "Econometrics Faced with the Needs of Macroeconomic Policy." *Econometrica*, 49, 1363–75.

Mallows, C. (1973). "Some Comments on Cp." *Technometrics*, 15, 661–75.

March, L., and M. Batty. (1975). "Generalized Measures of Information, Bayes' Likelihood Ratio and Jaynes' Formalism." *Environment and Planning B*, 2, 99–105.

Mariano, R. (1982). "Analytical Small–Sample Distribution Theory in Econometrics: The Simultaneous Equation Case." *International Economic Review*, 23, 503–33.

Marschak, J. (1939). "On Combining Market and Budget Data in Demand Studies: A Suggestion." *Econometrica*, 7, 332–35.

Martin, R. (1974). "On Spatial Dependence, Bias and the Use of First Spatial Differences in Regression Analysis." *Area*, 6, 185–94.

Martin, R. (1987). "Some Comments on Correction Techniques for Boundary Effects and Missing Value Techniques." *Geographical Analysis*, 19, 273–82.

Martin, R., and J. Oeppen. (1975). "The Identification of Regional Forecasting Models Using Space–Time Correlation Functions." *Transactions of the Institute of British Geographers*, 66, 95–118.

Matheron, G. (1971). *The Theory of Regionalized Variables*. Fontainebleau: Centre de Morphologie Mathematique.

Matula, D. and R. Sokal. (1980). "Properties of Gabriel Graphs Relevant to Geographic Variation Research and the Clustering of Points in the Plane." *Geographical Analysis*, 12, 205–222.

Mayer, T. (1975). "Selecting Economic Hyptotheses by Goodness of Fit." *Economic Journal*, 85, 877–83.

Mayer, T. (1980). "Economics as a Hard Science: Realistic Goal or Wishful Thinking." *Economic Inquiry*, 18, 165–78.

Mazodier, P., and A. Trognon. (1978). "Heteroskedasticity and Stratification in Error Components Models." *Annales de l'INSEE*, 30/31, 451–82.

McAleer, M. (1983). "Exact Tests of a Model Against Nonnested Alternatives." *Biometrika*, 70, 285–88.

McAleer, M. (1987). "Specification Tests of Separate Models: A Survey." In *Specification Analysis in the Linear Model*, edited by M. King and D. Giles, pp. 146–96. London: Routledge & Kegan Paul.

McAleer, M. and M. Pesaran. (1986). "Statistical Inference in Non–Nested Econometric Models." *Applied Mathematics and Computation*, 20, 271–311.

McCallum, B. (1976). "On Estimation Assuming Non–Existent Autocorrelation." *Australian Economic Papers*, 15, 119–27.

McCamley, F. (1973). "Testing for Spatial Autocorrelated Disturbances with Application to Relationships Estimated Using Missouri County Data." *Regional Science Perspectives*, 3, 89–104.

McLeish, D. (1975). "A Maximal Inequality and Dependent Strong Laws." *The Annals of Probability*, 3, 829–39.

Mead, R. (1967). "A Mathematical Model for the Estimation of Interplant Competition." *Biometrics*, 30, 295–307.

Meeks, S., and R. D'Agostino. (1983). "A Note on the Use of Confidence Limits Following Rejection of a Null Hypothesis." *The American Statistician*, 37, 134–6.

Miller, R. (1974). "The Jackknife – A Review." *Biometrika*, 61, 1–15.

Mizon, G., and D. Hendry. (1980). "An Empirical Application and Monte Carlo Analysis of Tests of Dynamic Specification." *Review of Economic Studies*, 47, 21–45.

Mizon, G., and J–F. Richard. (1986). "The Encompassing Principle and its Application to Testing Non–Nested Hypotheses." *Econometrica*, 54, 657–78.

Mood, A., F. Graybill, and D. Boes. (1974). *Introduction to the Theory of Statistics*. New York: McGraw Hill.

Moran, P. (1948). "The Interpretation of Statistical Maps." *Journal of the Royal Statistical Society B*, 10, 243–51.

Mosteller, F., and J. Tukey. (1977). *Data Analysis and Regression. A Second Course in Statistics*. Reading, MA: Addison–Wesley.

Nakamura, A., and M. Nakamura. (1978). "On the Impact of the Tests for Serial Correlation upon the Test of Significance for the Regression Coefficient." *Journal of Econometrics*, 7, 199–210.

Nerlove, M. (1971). "Further Evidence on the Estimation of Dynamic Economic Relations From a Time Series of Cross Sections." *Econometrica*, 39, 359–82.

Neudecker, H. (1969). "Some Theorems on Matrix Differentiation with Special Reference to Kronecker Matrix Products." *Journal of the American Statistical Association*, 64, 953–63.

Newey, W. (1985a). "Maximum Likelihood Specification Testing and Conditional Moment Tests." *Econometrica*, 53, 1047–70.

Newey, W. (1985b). "Generalized Methods of Moments Specification Testing." *Journal of Econometrics*, 29, 229–56.

Newey, W. and K. West. (1987). "A Simple, Positive Semi–Definite, Heteroskedasticity and Autocorrelation Consistent Covariance Matrix." *Econometrica*, 55, 703–8.

Nicholls, D. and A. Pagan. (1983). "Heteroskedasticity in Models with Lagged Dependent Variables." *Econometrica*, 51, 1233–42.

Nijkamp, P., H. Leitner, and N. Wrigley. (1985). *Measuring the Unmeasurable*. Dordrecht: Martinus Nijhoff.

Nipper, J. and U. Streit. (1982). "A Comparative Study of Some Stochastic Methods and Autoprojective Models for Spatial Processes." *Environment and Planning A*, 14, 1211–1231.

Oberhoffer, W., and J. Kmenta. (1974). "A General Procedure for Obtaining Maximum Likelihood Estimates in Generalized Regression Models." *Econometrica*, 42, 579–90.

Odland, J. (1978). "Prior Information in Spatial Analysis." *Environment and Planning A*, 10, 51–70.

Openshaw, S. (1984). "Ecological Fallacies and the Analysis of Areal Census Data." *Environment and Planning A*, 16, 17–31.

Openshaw, S., and P. Taylor. (1979). "A Million or so Correlation Coefficients: Three Experiments on the Modifiable Areal Unit Problem." In *Statistical Applications in the Spatial Sciences*, edited by N. Wrigley, pp. 127–44. Pion: London.

Openshaw, S., and P. Taylor. (1981). "The Modifiable Unit Problem." In *Quantitative Geography, a British View*, edited by N. Wrigley and R. Bennett, pp. 60–69. London: Routledge and Kegan.

Ord, J. (1975). "Estimation Methods for Models of Spatial Interaction." *Journal of the American Statistical Association*, 70, 120–26.

Ord, J. (1981). "Towards a Theory of Spatial Statistics: A Comment." *Geographical Analysis*, 13, 86–93.

Paelinck, J. (1982). "Operational Spatial Analysis." *Papers, Regional Science Association*, 50, 1–7.

Paelinck, J. and L. Klaassen. (1979). *Spatial Econometrics*. Farnborough: Saxon House.

Paelinck, J., J. Ancot, and J. Kuipers. (1982). *Formal Spatial Economic Analysis*. Aldershot: Gower Publishing.

Pagan, A. (1980). "Some Identification and Estimation Results for Regression Models with Stochastically Varying Coefficients." *Journal of Econometrics*, 13, 341–63.

Pagan, A., and A. Hall. (1983). "Diagnostic Tests as Residual Analysis." *Econometric Reviews*, 2, 159–218.

Parks, R. (1967). "Efficient Estimation of a System of Regression Equations when Disturbances are both Serially and Contemporaneously Correlated." *Journal of the American Statistical Association*, 62, 500–9.

Peach, J., and J. Webb. (1983). "Randomly Specified Macroeconomic Models: Some Implications for Model Selection." *Journal of Economic Issues*, 17, 697–714.

Pesaran, M. (1974). "On the General Problem of Model Selection." *Review of Economic Studies*, 41, 153–71.

Pesaran, M. (1981). "Pitfalls of Testing Non–Nested Hypotheses by the Lagrange Multiplier Method." *Journal of Econometrics*, 17, 323–31.

Pesaran, M. (1982a). "On the Comprehensive Method of Testing Non–Nested Regression Models." *Journal of Econometrics*, 18, 263–74.

Pesaran, M. (1982b). "Comparison of Local Power of Alternative Tests of Non–Nested Regression Models." *Econometrica*, 50, 1287–1305.

Pesaran, M. and A. Deaton. (1978). "Testing Non–Nested Nonlinear Regression Models." *Econometrica*, 46, 677–94.

Pfeifer, P., and S. Deutsch. (1980a). "A Three–Stage Iterative Procedure for Space–Time Modeling." *Technometrics*, 22, 35–47.

Pfeifer, P., and S. Deutsch. (1980b). "Identification and Interpretation of First Order Space–Time ARMA Models." *Technometrics*, 22, 397–408.

Pfeiffer, P. (1978). *Concepts of Probability Theory*. New York: Dover.

Phillips, P. (1977). "A General Theorem in the Theory of Asymptotic Expansions as Approximations to the Finite Sample Distributions of Econometric Estimators." *Econometrica*, 45, 1517–34.

Phillips, P. (1980). "The Exact Distribution of Instrumental Variable Estimators in an Equation Containing N+1 Endogenous Variables." *Econometrica*, 48, 861–78.

Phillips, P. (1982). "Best Uniform and Modified Pade Approximants to Probability Densities in Econometrics." In *Advances in Econometrics*, edited by W. Hildenbrand, pp. 123–67. London: Cambridge University Press.

Phillips, P. (1985). "The Exact Distribution of the SUR Estimator." *Econometrica*, 53, 745–56.

Pindyck, R. and D. Rubinfeld. (1981). *Econometric Models and Economic Forecasts*. New York: McGraw–Hill.

Pocock, S., D. Cook, and A. Shaper. (1982). "Analysing Geographic Variation in Cardiovascular Mortality: Methods and Results." *Journal of the Royal Statistical Society A*, 145, 313–341.

Pollock, D. (1979). *The Algebra of Econometrics*. New York: John Wiley.

Pool, I. and M. Kochen. (1978). "Contacts and Influence." *Social Networks*, 1, 5–51.

Prucha, I. (1984). "On the Asymptotic Efficiency of Feasible Aitken Estimators for Seemingly Unrelated Regression Models with Error Components." *Econometrica*, 52, 203–7.

Quandt, R. (1958). "The Estimation of the Parameters of a Linear Regression System Obeying Two Separate Regimes." *Journal of the American Statistical Association*, 53, 873–80.

Quandt, R. (1972). "A New Approach to Estimating Switching Regressions." *Journal of the American Statistical Association*, 67, 206–10.

Quandt, R. (1974). "A Comparison of Methods for Testing Nonnested Hypotheses." *Review of Economics and Statistics*, 56, 92–9.

Quandt, R. (1981). "Autocorrelated Errors in Simple Disequilibrium Models." *Economics Letters*, 7, 55–61.

Quandt, R. (1982). "Econometric Disequilibrium Models." Econometric Reviews, 1, 1–63.

Quandt, R. and J. Ramsey. (1978). "Estimating Mixtures of Normal Distributions and Switching Regressions." Journal of the American Statistical Association, 73, 730–52.

Raj, B., and A. Ullah. (1981). *Econometrics, A Varying Coefficients Approach*. New York: St Martins Press.

Ramsey, J. (1974). "Classical Model Selection Through Specification Tests." In *Frontiers in Econometrics*, edited by P. Zarembka, pp. 13–47. New York: Academic Press.

Ripley, B. (1981). *Spatial Statistics*. New York: Wiley.

Robinson, P. (1987). "Asymptotically Efficient Estimation in the Presence of Heteroskedasticity of Unknown Form." *Econometrica*, 55, 875−91.

Rogers, G. (1980). *Matrix Derivatives*. New York: Marcel Dekker.

Rosenberg, B. (1973). "A Survey of Stochastic Parameter Regression." *Annals of Economic and Social Measurement*, 2/4, 381−97.

Rosenblatt, M. (1956). "A Central Limit Theorem and a Strong Mixing Condition." *Proceedings of the National Academy of Sciences*, 42, 43−47.

Rothenberg, T. (1982). "Comparing Alternative Asymptotically Equivalent Tests." In *Advances in Econometrics*, edited by W. Hildenbrand, pp. 255−62. London: Cambridge University Press.

Rothenberg, T. (1984a). "Approximate Normality of Generalized Least Squares Estimates." *Econometrica*, 52, 811−25.

Rothenberg, T. (1984b). "Hypothesis Testing in Linear Models when the Error Covariance Matrix is Nonscalar." *Econometrica*, 52, 827−42.

Ruud, P. (1984). "Tests of Specification in Econometrics." *Econometric Reviews*, 3, 211−42.

Sargan, J. (1958). "The Estimation of Economic Relationships Using Instrumental Variables." *Econometrica*, 26, 393−415.

Sargan, J. (1976). "Econometric Estimators and the Edgeworth Approximation." *Econometrica*, 44, 421−48.

Sargan, J. (1980). "Some Tests of Dynamic Specification for a Single Equation." *Econometrica*, 48, 879−97.

Savin, N. (1976). "Conflict Among Testing Procedures in a Linear Regression Model with Autoregressive Disturbances." *Econometrica*, 44, 1303−15.

Savin, N. (1980). "The Bonferroni and Scheffe Multiple Comparison Procedures." *Review of Economic Studies*, 47, 255−74.

Savin, N., and K. White. (1978). "Estimation and Testing for Functional Form and Autocorrelation: A Simultaneous Approach." *Journal of Econometrics*, 8, 1−12.

Sawa, T. (1978). "Information Criteria for Discriminating among Alternative Regression Models." *Econometrica*, 46, 1273−91.

Schmidt, P. (1977). "Estimation of Seemingly Unrelated Regressions with Unequal Numbers of Observations." *Journal of Econometrics*, 5, 365−77.

Schmidt, P. and R. Sickles. (1977). "Some Further Evidence on the Use of the Chow Test under Heteroskedasticity." *Econometrica*, 45, 1293−8.

Schoemaker, P. (1982). "The Expected Utility Model: Its Variants, Purposes, Evidence and Limitations." *Journal of Economic Literature*, 20, 529−63.

Schulze, P. (1977). "Testing for Intraregional Economic Homogeneity: A Comment." *Journal of Regional Science*, 17, 473−77.

Schulze, P. (1987). "Once Again: Testing for Regional Homogeneity." *Journal of Regional Science*, 27, 129−33.

Scott, A. and D. Holt. (1982). "The Effect of Two−Stage Sampling on Ordinary Least Squares Methods." *Journal of the American Statistical Association*, 77, 848−54.

Schwallie, D. (1982). "Unconstrained Maximum Likelihood Estimation of Contemporaneous Covariances." *Economics Letters*, 9, 359−64.

Schwartz, G. (1978). "Estimating the Dimension of a Model." *Annals of Statistics*, 6, 461−4.

Sen, A., and S. Soot. (1977). "Rank Test for Spatial Correlation." *Environment and Planning A*, 9, 897−903.

Serfling, R. (1980). *Approximation Theorems of Mathematical Statistics*. New York: John Wiley.

Shapiro, H. (1973). "Is Verification Possible? The Evaluation of Large Econometric Models." *American Journal of Agricultural Economics*, 55, 250−8.

Sheppard, E. (1976). "Entropy, Theory Construction and Spatial Analysis." *Environment and Planning A*, 8, 741—52.

Silvey, S. (1961). "A Note on Maximum—Likelihood in the Case of Dependent Random Variables." *Journal of the Royal Statistical Society B*, 23, 444—52.

Sims, C. (1980). "Macroeconomics and Reality." *Econometrica*, 48, 1—48.

Singh, B., A. Nagar, N. Choudry, B. Raj. (1976). "On the Estimation of Structural Change: A Generalization of the Random Coefficients Regression Model." *International Economic Review*, 17, 340—61.

Singh, B., and A. Ullah. (1974). "Estimation of Seemingly Unrelated Regressions with Random Coefficients." *Journal of the American Statistical Association*, 69, 191—5.

Smith, D., and B. Hutchinson. (1981). "Goodness—of—Fit Statistics for Trip Distribution Models." *Transportation Research A*, 15, 295—303.

Smith, P., and S. Choi. (1982). "Simple Tests to Compare Two Dependent Regression Lines." *Technometrics*, 24, 123—6.

Smith, T. (1980). "A Central Limit Theorem for Spatial Samples." *Geographical Analysis*, 12, 299—324.

Snickars, F., and T. Weibull. (1977). "A Minimum Information Principle: Theory and Practice." *Regional Science and Urban Economics*, 7, 137—68.

Spencer, D. (1979). "Estimation of a Dynamic System of Seemingly Unrelated Regressions with Autoregressive Disturbances." *Journal of Econometrics*, 10, 227—41.

Spencer, D. and K. Berk. (1981). "A Limited Information Specification Test." *Econometrica*, 49, 1079—85.

Srivastava, V., and T. Dwivedi. (1979). "Estimation of Seemingly Unrelated Regression Equations. A Brief Survey." *Journal of Econometrics*, 10, 15—32.

Stetzer, F. (1982a). "The Analysis of Spatial Parameter Variation with Jackknifed Parameters." *Journal of Regional Science*, 22, 177—88.

Stetzer, F. (1982b). "Specifying Weights in Spatial Forecasting Models: The Results of Some Experiments." *Environment and Planning A*, 14, 571—84.

Stone, M. (1974). "Cross—Validatory Choice and Assessment of Statistical Predictions." *Journal of the Royal Statistical Society B*, 36, 111—33.

Summerfield, M. (1983). "Populations, Samples and Statistical Inference in Geography." *The Professional Geographer*, 35, 143—49.

Swamy, P. (1971). *Statistical Inference in Random Coefficient Regression Models*. New York: Springer Verlag.

Swamy, P. (1974). "Linear Models with Random Coefficients." In *Frontiers in Econometrics*, edited by P. Zarembka, pp. 143—68. New York: Academic Press.

Swamy, P., R. Conway, and P. von zur Muehlen. (1985). "The Foundations of Econometrics — are there any? *Econometric Reviews*, 4, 1—61.

Szyrmer, J. (1985). "Measuring Connectedness of Input—Output Models: 1. Survey of the Measures." *Environment and Planning A*, 17, 1591—1612.

Tauchen, G. (1985). "Diagnostic Testing and Evaluation of Maximum Likelihood Models." *Journal of Econometrics*, 30, 415—43.

Taylor, C. (1982a). "Econometric Modeling of Urban and Other Substate Areas: An Analysis of Alternative Methodologies." *Regional Science and Urban Economics*, 12, 425—48.

Taylor, C. (1982b). "Regional Econometric Model Comparisons: What Do They Mean." *The Annals of Regional Science*, 18, 1—15.

Taylor, W. (1983). "On the Relevance of Finite Sample Distribution Theory." *Econometric Reviews*, 2, 1—39.

Theil, H. (1971). *Principles of Econometrics*. New York: John Wiley.

Thomas, R. (1981). *Information Statistics in Geography*. Norwich: Geo—Abstracts.

Thompson, M. (1978a). "Selection of Variables in Multiple Regression: Part I. A Review and Evaluation." *International Statistical Review*, 46, 1–19.

Thompson, M. (1978b). "Selection of Variables in Multiple Regression: Part II. Chosen Procedures, Computations and Examples." *International Statistical Review*, 46, 129–46.

Thursby, J. (1981). "A Test Strategy for Discriminating Between Autocorrelation and Misspecification in Regression Analysis." *The Review of Economics and Statistics*, 63, 117–23.

Thursby, J. (1982). "Misspecification, Heteroskedasticity, and the Chow and Goldfeld–Quandt Tests." *The Review of Economics and Statistics*, 64, 314–21.

Tinbergen, J. (1940). "On a Method of Statistical Business Research: A Reply." *Economic Journal*, 50, 141–54.

Tinbergen, J. (1942). "Critical Remarks on some Business Cycle Theories." *Econometrica*, 10, 129–46.

Tinkler, K. (1972). "The Physical Interpretation of Eigenfunctions of Dichotomous Matrices." *Transactions, Institute of British Geographers*, 55, 17–46.

Tjostheim, D. (1978). "A Measure of Association for Spatial Variables." *Biometrika*, 65, 109–114.

Tobler, W. (1979). "Cellular Geography." In *Philosophy in Geography*, edited by S. Gale and G. Olsson, pp. 379–86. Dordrecht: Reidel.

Toyoda, T. (1974). "The Use of the Chow Test under Heteroskedasticity." *Econometrica*, 42, 601–8.

Upton, G. and B. Fingleton. (1985). *Spatial Data Analysis by Example*. New York: Wiley.

Verbon, H. (1980a). "Testing for Heteroscedasticity in a Model of Seemingly Unrelated Regression Equations with Variance Components (SUREVC)." *Economics Letters*, 5, 149–53.

Verbon, H. (1980b). "Maximum Likelihood Estimation of a Labour Demand System. An Application of a Model of Seemingly Unrelated Regression Equations with the Regression Errors Composed of Two Components." *Statistica Neerlandica*, 34, 33–48.

Vinod, H., and A. Ullah. (1981). *Recent Advances in Regression Methods*. New York: Marcel Dekker.

Walker, A. (1967). "Some Tests of Separate Families of Hypotheses in Time Series Analysis." *Biometrika*, 54, 39–68.

Wallace, T. (1977). "Pre–Test Estimation in Regression: A Survey." *American Journal of Agricultural Economics*, 59, 431–43.

Wallace, T., and A. Hussain. (1969). "The Use of Error Components Models in Combining Cross Section with Time Series Data." *Econometrica*, 37, 55–72.

Wallis, K. (1967). "Lagged Dependent Variables and Serially Correlated Errors: a Reappraisal of Three–Pass Least Squares." *Review of Economics and Statistics*, 51, 555–67.

Wallsten, T., and D. Budescu. (1983). "Encoding Subjective Probabilities: A Psychological and Psychometric Review." *Management Science*, 29, 151–73.

Wang, G., M. Hidiroglou, and W. Fuller. (1980). "Estimation of Seemingly Unrelated Regression with Lagged Dependent Variables and Autocorrelated Errors." *Journal of Statistical Computation and Simulation*, 10, 133–46.

Wansbeek, T., and A. Kapteyn. (1982). "A Class of Decompositions of the Variance–Covariance Matrix of a Generalized Error Components Model." *Econometrica*, 50, 713–24.

Wartenberg, D. (1985). "Multivariate Spatial Correlation: A method for Exploratory Geographical Analysis." *Geographical Analysis*, 17, 263–83.

Watt, P. (1979). "Tests of Equality Between Sets of Coefficients in Two Linear Regressions When Disturbance Variances are Unequal: Some Small Sample Properties." *The Manchester School of Economic and Social Studies*, 47, 391–6.

Weibull, J. (1976). "An Axiomatic Approach to the Measurement of Accessibility." *Regional Science and Urban Economics*, 6, 357−79.

Welsch, R. (1980). "Regression Sensitivity Analysis and Bounded−Influence Estimation." In *Evaluation of Econometric Models*, edited by J. Kmenta and J. Ramsey, pp. 153−67. New York: Academic Press.

White, D., M. Burton and M. Dow. (1981). "Sexual Division of Labor in African Agriculture: A Network Autocorrelation Analysis." *American Anthropologist*, 84, 824−49.

White, E. and G. Hewings. (1982). "Space−Time Employment Modeling: Some Results Using Seemingly Unrelated Regression Estimators." *Journal of Regional Science*, 22, 283−302.

White, H. (1980). "A Heteroskedastic−Consistent Covariance Matrix Estimator and a Direct Test for Heteroskedasticity." *Econometrica*, 48, 817−38.

White, H. (1982a). "Maximum Likelihood Estimation of Misspecified Models." *Econometrica*, 50, 1−25.

White, H. (1982b). "Regularity Conditions for Cox's Test of Non−Nested Hypotheses." *Journal of Econometrics*, 19, 301−18.

White, H. (1984). *Asymptotic Theory for Econometricians*. New York: Academic Press.

White, H., and I. Domowitz. (1984). "Nonlinear Regression with Dependent Observations." *Econometrica*, 52, 143−61.

Whittle, P. (1954). "On Stationary Processes in the Plane." *Biometrika*, 41, 434−49.

Wilkinson, J., and C. Reinsch. (1971). *Linear Algebra, Handbook for Automatic Computation, Vol. 2*. Berlin: Springer Verlag.

Wilks, S. (1962). *Mathematical Statistics*. New York: Wiley.

Wilson, A. (1970). *Entropy in Urban and Regional Modeling*. London: Pion.

Wrigley, N. (1979). *Statistical Applications in the Spatial Sciences*. London: Pion.

Wu, D−M. (1973). "Alternative Tests of Independence Between Stochastic Regressors and Disturbances." *Econometrica*, 41, 733−40.

Wu, D−M. (1974). "Alternative Tests of Independence Between Stochastic Regressors and Disturbances: Finite Sample Results." *Econometrica*, 42, 529−46.

Zellner, A. (1962). "An Efficient Method of Estimating Seemingly Unrelated Regressions and Tests of Aggregation Bias." *Journal of the American Statistical Association*, 57, 348−68.

Zellner, A. (1971). *An Introduction to Bayesian Inference in Econometrics*. New York: John Wiley.

Zellner, A. (1978). "Jeffreys−Bayes Posterior Odds Ratio and the Akaike Information Criterion for Discriminating Between Models." *Economics Letters*, 1, 337−42.

Zellner, A. (1979). "Statistical Analysis of Econometric Models." *Journal of the American Statistical Association*, 74, 628−43.

Zellner, A. (1980). *Bayesian Analysis in Econometrics and Statistics, Essays in Honor of Harold Jeffreys*. Amsterdam: North Holland.

Zellner, A. (1984). *Basic Issues in Econometrics*. Chicago: University of Chicago Press.

Zellner, A. (1985). "Bayesian Econometrics." *Econometrica*, 53, 253−69.

Zellner, A., and F. Palm. (1974). "Time Series Analysis and Simultaneous Equation Econometric Models." *Journal of Econometrics*, 2, 17−54.

Zellner, A., and G. Tiao. (1964). "Bayesian Analysis of the Regression Model with Autocorrelated Errors." *Journal of the American Statistical Association*, 59, 763−78.

Ziemer, R. (1984). "Reporting Econometric Results: Believe it or Not? *Land Economics*, 60, 122−7.

INDEX